THE CRAFTER'S
GUIDE TO GLUES

OTHER BOOKS AVAILABLE FROM CHILTON

Robbie Fanning, Series Editor

Contemporary Quilting

All Quilt Blocks Are Not Square, by Debra Wagner

Barbara Johannah's Crystal Piecing

The Complete Book of Machine Quilting, Second Edition, by Robbie and Tony Fanning

Contemporary Quilting Techniques, by Pat Cairns

Creative Triangles for Quilters, by Janet B. Elwin

Fast Patch, by Anita Hallock

Fourteen Easy Baby Quilts, by Margaret Dittman

Machine-Quilted Jackets, Vests, and Coats, by Nancy Moore

Pictorial Quilts, by Carolyn Vosburg Hall

Precision Pieced Quilts Using the Foundation Method, by Jane Hall and Dixie Haywood

Quick-Quilted Home Decor with Your Bernina, by Jackie Dodson

Quick-Quilted Home Decor with Your Sewing Machine, by Jackie Dodson

The Quilter's Guide to Rotary Cutting, by Donna Poster

Scrap Quilts Using Fast Patch, by Anita Hallock

Shirley Botsford's Daddy's Ties

Speed-Cut Quilts, by Donna Poster

Stitch 'n' Quilt, by Kathleen Eaton

Super Simple Quilts, by Kathleen Eaton

Teach Yourself Machine Piecing and Quilting, by Debra Wagner

Three-Dimensional Appliqué, by Jodie Davis

Three-Dimensional Pieced Quilts, by Jodie Davis

Craft Kaleidoscope

Creating and Crafting Dolls, by Eloise Piper and Mary Dilligan

Fabric Crafts and Other Fun with Kids, by Susan Parker Beck and Charlou Lunsford

Fabric Painting Made Easy, by Nancy Ward

Jane Asher's Costume Book

Quick and Easy Ways with Ribbon, by Ceci Johnson

Learn Bearmaking, by Judi Maddigan

Soft Toys for Babies, by Judi Maddigan

Stamping Made Easy, by Nancy Ward

Too Hot To Handle? Potholders and How to Make Them, by Doris L. Hoover

Creative Machine Arts

ABCs of Serging, by Tammy Young and Lori Bottom

The Button Lover's Book, by Marilyn Green

Claire Shaeffer's Fabric Sewing Guide

The Complete Book of Machine Embroidery, by Robbie and Tony Fanning

Creative Nurseries Illustrated, by Debra Terry and Juli Plooster

Distinctive Serger Gifts and Crafts, by Naomi Baker and Tammy Young

Friendship Quilts by Hand and Machine, by Carolyn Vosburg Hall

Gail Brown's All-New Instant Interiors

Hold It! How to Sew Bags, Totes, Duffels, Pouches, and More, by Nancy Restuccia

How to Make Soft Jewelry, by Jackie Dodson

Innovative Serging, by Gail Brown and Tammy Young

Innovative Sewing, by Gail Brown and Tammy Young

Jan Saunders' Wardrobe Quick-Fixes

The New Creative Serging Illustrated, by Pati Palmer, Gail Brown, and Sue Green

Petite Pizzazz, by Barb Griffin

Putting on the Glitz, by Sandra L. Hatch and Ann Boyce

Quick Napkin Creations, by Gail Brown

Second Stitches: Recycle as You Sew, by Susan Parker

Serge a Simple Project, by Tammy Young and Naomi Baker

Serge Something Super for Your Kids, by Cindy Cummins

Serged Garments in Minutes, by Tammy Young and Naomi Baker

Sew Any Patch Pocket, by Claire Shaeffer

Sew Any Set-In Pocket, by Claire Shaeffer

Sew Sensational Gifts, by Naomi Baker and Tammy Young

Sewing and Collecting Vintage Fashions, by Eileen MacIntosh

Simply Serge Any Fabric, by Naomi Baker and Tammy Young

Soft Gardens: Make Flowers with Your Sewing Machine, by Yvonne Perez-Collins

The Stretch & Sew Guide to Sewing Knits, by Ann Person

Twenty Easy Machine-Made Rugs, by Jackie Dodson

Know Your Sewing Machine Series, by Jackie Dodson

Know Your Bernina, second edition

Know Your Brother, with Jane Warnick

Know Your New Home, with Judi Cull and Vicki Lyn Hastings

Know Your Pfaff, with Audrey Griese

Know Your Sewing Machine

Know Your Singer

Know Your Viking, with Jan Saunders

Know Your White, with Jan Saunders

Know Your Serger Series, by Tammy Young and Naomi Baker

Know Your baby lock

Know Your Serger

Know Your White Superlock

StarWear

Dazzle, by Linda Fry Kenzle

Embellishments, by Linda Fry Kenzle

Make It Your Own, by Lori Bottom and Ronda Chaney

Mary Mulari's Garments with Style

Pattern-Free Fashions, by Mary Lee Trees Cole

Shirley Adams' Belt Bazaar

Sweatshirts with Style, by Mary Mulari

Teach Yourself to Sew Better, by Jan Saunders

A Step-by-Step Guide to Your Bernina

A Step-by-Step Guide to Your New Home

A Step-by-Step Guide to Your Sewing Machine

A Step-by-Step Guide to Your Viking

THE CRAFTER'S
GUIDE TO GLUES

Tammy Young

KALEIDO
CRAFT
·SCOPE·

Chilton Book Company
Radnor, Pennsylvania

Published in Radnor, Pennsylvania 19089, by Chilton Book Company

The author and publisher have made every effort to ensure that all information and instructions given in this book are accurate and safe, but they
accept no responsibility or liability for any injury, damage, or loss resulting
from misuse or abuse of any product. The user is responsible for reading
and following manufacturer's directions on all products, especially those
which are flammable or contain dangerous solvents. The reader is also
alerted by **Caution** notices throughout the instructions in this book when
special care is required.

Designed by General Graphic Services
Manufactured in Mexico

Library of Congress Cataloging in Publication Data
Tammy, Young.
    The crafter's guide to glues / Tammy Young.
       p.   cm.—(Craft kaleidoscope)
    Includes bibliographical references and index.
    ISBN 0-8019-8611-7 (pbk.)
    1. Handicraft—Equipment and supplies.   2. Glue.   3. Adhesives.
I. Title. II. Series.
TT153.7. Y68   1996   95-31155
745.5-028—dc20   CIP

2 3 4 5 6 7 8 9 0   4 3 2 1 0 9 8 7

Contents

Preface

When I was growing up, I never considered myself a "crafty" type. I did go through a sticker phase and an Indian beading period as a young child, but sewing and then serging have been the major focus of my hobbyist activities.

Why, then, did I decide to write a book on crafting with glues? Because every time I wanted to glue something to something else, I didn't know which glue to use or even where to turn for advice.

Since I proposed this book to Chilton (my fourteenth with them to date) and began the research, I have never had more comments, suggestions, biased recommendations, misinformation, good advice, and equal amounts of discouragement and encouragement. One knowledgeable industry professional said, "But you're trying to bring order to an industry that has no order."

Naive and undaunted, I plunged ahead. After I asked many more questions and did more and more testing, the glues and adhesives began to sort themselves into reasonable categories for crafters (although there's still some overlap). My twofold intentions were: first, to put this complicated subject into an order I could understand; and second, to write about it so that you could, too. Since then, I have had various reactions from those who heard that I was working on the book—everything from "Is there really enough for a whole book?" through "Great! It's about time." to "How in the world can you possibly cover it all?" Those making the latter comment were the most "in the know." It's been impossible to include every crafting option and every available product, but I tried to incorporate as many alternatives as the space allowed.

A few tips for sewing and quilting enthusiasts are detailed, but this book places the most emphasis on true crafting techniques with fast and simple methods for constructing a wide variety of projects (each project includes step-by-step instructions). You'll learn which glue to use and how to use it for everything from simple kids' crafts, to fabric and embellishment techniques, to technical specialty applications. One chapter features the popular hot glues.

The beauty of crafting with glue is that anyone can do it and there are a wide variety of project choices for all ages, tastes, and abilities. Now, thanks to working on this book, I'm energized to explore all sorts of crafting projects. With the wide range of adhesives available to us today (and knowing something about how to use them), glue crafting can be fun, fast, and very rewarding.

After reading this book, I hope you'll feel just like I do—"craftier" than ever! Enjoy!

Tammy Young

Acknowledgements

This book could not have been completed without the contribution of so many helpful people, all of whom have my sincere gratitude.

First, many thanks to the numerous industry companies listed in this book who gave me information, tips, and product samples for testing. I want to express special appreciation to those individuals who were most generous with their in-depth advice and support: Mike Assile of Beacon Chemical Co., Richard Fragiacomo of Bond, John Anderson of Elmer's Laboratory at Borden, Clotilde of Clotilde, Inc., Sharion Cox of Cox Co., Shirley Penman of Creative Potential, and Miriam Olson of *Profitable Craft Merchandising (PCM)* magazine.

A big thanks also to fellow Chilton authors Nancy Ward, Yvonne Perez-Collins, and Gail Brown, who shared so much of their creativity and knowledge as I was researching and writing. My appreciation also goes to all of the Chilton authors who contributed special tips featured throughout the book, and especially to Naomi Baker, who also listened to much of my frustration as I struggled to make sense of the subject.

Others deserving special thanks for their help during my research and writing include: Karen Dillon (who tested most of the fabric and embellishment glues), Leslie Wood (who did part of the initial research for the book), and Sarah Corrales (who cheerfully helped with several testing projects whenever she came to visit). My gratitude also to Jolly Michel, Maria Filosa, Kathy Nowicki, and Susan Jones of Jones Tones for their generous contributions and artistic talent.

Another huge thanks to Chris Hansen who not only illustrated this book in his incomparable style but also contributed a wealth of information from personal experience. And last but not least, thanks to the entire Chilton team who worked so hard to edit and produce the book.

The following trademarked and registered names appear in this book:

3M: Metal & Ceramic Adhesive, Super Glue Gel, Super Strength Adhesive, Wood Glue

Accent: Tacky Glue

Activa: Mighty Tacky

Adhesive Technologies: Crafty Magic Melt, Floral Pro, Little Dipper, Stik-A-Roo

Aleene's: 3-D Foiling Glue, Fabric Stiffener and Fabric Draping Liquid, Fine Line Syringe, Flexible Stretchable Fabric Glue, Hot Stitch, Hot Stitch Glue, Hot Stitch Ultra Hold, Jewel-It, Leather Glue, Liquid Fusible Web, No-Sew, Ok to Wash-It, Paper Napkin Applique Glue, Professional Wood Glue, Reverse Collage Glue, Right-On, Stop Fraying, Tack-It Over & Over, "Tacky" Glue, Thick Designer Tacky Glue, Thin Bodied Tacky Glue, White Glue

Anita's: Découpage Glue, Foil Adhesive

Attack

Back Street: Border Patrol

Barge: All-Purpose Cement

Beacon: Fabri Tac, Gem Tac, Liquid Laminate, Liqui Fuse, Stiffen Stuff

Best-Test: Grade School Glue, Pik-Up

Blair: Stencil-Stik

Bond: 484:Tacky, 527 Multi Purpose Cement, Crackle & Age, E-Z-Tak, Fabric Glue, Floral Foam Bond, Get Set, Heat & Seal, Instant Grrrip, Jewel Glue, No Fray, Squeezable Tacky, Stik, Victory 1991, Victory Fabric Glue, Victory Fast Grab, Victory Household Cement, Wash It Again & Again

Borden: Craft Bond I, Krafty Glue LSA, Krazy Glue, Krazy Glue Gel

Brohman: No-Fray

Convenience Products: Arrange-It, Clean-It, Set-It

Craftmate

Crayola: Art & Craft Glue

Creatively Yours: Clear Silicone, Crafter's Cement, Fabric Plus, Jewelry & More, Stencil Adhesive, Quick Set Epoxy Pouches

Creative Potential: Sticky Stuff

Crescent: Perfect Mount Film, Sand Expressions Film

Dab Stic

DAP: Dow Corning Brand Silicone

Darice: Plastic Canvas

Dcor: Prestik

Delta: Jewel Glue, Naplique Paper Applique, Quik 'n Tacky, Second Impressions, Sheer impressions, Sobo, Stencil Magic, Stitchless, Velverette, Woodwiz

Devcon: 2-Ton Epoxy, 5 Minute Epoxy, Clear Silicone Rubber, Duco, Duco Cement, Duco Plastic & Model Cement, Duco Stik-Tak, Gripwood White Glue, Plastic Welder, Super Glue Gel, Weldit Cement

Distlefink Designs: Hot Tape

Dritz: Fray Check, Hem N Trim, Insta Tack, iron-off, Liquid Stitch, Stitch Witchery, Stitch Witchery Plus With Grid

Duncan: Easy Foil Medium

Duro: Clear Silicone Sealer, Crystal Clear, Household Cement, Quick Gel, Master Mend, Stick-With-It

Eberhard Faber: UHU, UHU Bond-All, UHU Fimo Glue, UHU HOLDiT, UHU Liquid Glue Pen, UHU Stic Color

Eclectic: Amazing Goop, Amazing Goop Adhesive Products, Crafter's Goop, E-6000, Household Goop, Tube Gripper

Elmer's (Borden): Carpenter's Wood Glue, Elmer's-Tack Adhesive Putty, Epoxy, Glue-All, GluColors, Household Cement, Model & Hobby Cement, School Glue Gel, School Glue Stick, Stix-All, Super-Fast Epoxy, Weather-Tite Wood Glue

Emson: ML2000

Evans: Banner Wheat Paste

Fabric Mender Magic

Faultless: Bead Easy, Bead Easy Re-Apply, Clear Top Coat Adhesive, Fabricraft Solutions, Instant Microwave Stiffener

Flora Craft: Clear Glue

FPC: Surebonder, Surebonder Glue Skillet

Franklin International: Hide Glue, Home, Shop & Craft Glue, Titebond, Titebond II, White Glue

Friendly Plastic

Garrett Wade: 202GF Gap Filling Glue, Yellow Woodworkers Glue, Slo-Set Glue, Special Laminating Glue

Gemtool

Germanow-Simon: G-S Hypo-Tube Cement

Gesso

Glitz, Inc.: Glitz-it!

Golden Harvest

Gonicoll: Stick and Peel Glue

Gorilla Glue

Halcraft USA: Hal-Tech 2001

H. B. Fuller: Pow'r Stix

Hot-Fix

HTC: Stitch Witchery, Stitch Witchery Plus With Grid, Trans Web

Jenny Jem's: Confetti Glue

Itoya: O'Glue

Jones Tones: Plexi 400 Stretch Adhesive

Jurgen: Fabric Drape & Lace Stiffener, Jewel & Fabric Glue, Press n Peel, Tacky Craft Glue

J.W. Etc.'s: First Step

Kony Bond: 2 in 1 Glue Marker, Easy Paper Glue, Multi-Purpose White Glue, School Glue Marker, Super Glue Gel

Krylon: Easy Tack

Lakeside Plastics: Tri-Tix

LePage's: Arts & Crafts Glue, Clear Glue, Hot Tropicals, Original Glue, Original Strength Wood Glue, White Glue

Loctite: Depend II, Desk Set, Desk Set Ball Point Glue Pen, Desk Set Clear Gel Adhesive, Desk Set Clear Gel Glue, Desk Set Put Ups, Desk Set Safe & Easy White Glue, Gelmatic, House Works, Poxy Pouches, Quicktite, Quicktite Super Glue Gel, Stick 'n Seal, Wood Worx

Lomey: FloraLock Stem Adhesive

Lycra

Magic American: Fabric Mender Magic, Goo Gone

Magic Glue Wand

Mark Enterprises: Mrs. Glue For Glitter & More

Modern Miltex: Goo Loo Styrofoam Craft Adhesive

National Artcraft: Narco Glass Hold

Pacer: Zap A-Dap-A Goo

Pellon: Heavy-Duty Wonder-Under, Wonder-Under, Wonder-Web

Pentel: Roll'n Glue

Plaid: Fabric Stiffener, Glu 'N Wash, Jewel Glue, Liquid Beads, Liquid Beads Dimensional Bond, Mod Podge, Petal Porcelain, Press & Peel Foil, Royal Coat, Stikit Again & Again, Tacky Glue

Real Glass: Laminating Glue, Liquid Leading

Repcon International: Create-a-Valance, Lite-Line, Tissue Box Topper

Ross: Art Paste, Epoxy Adhesive, Fast Set Epoxy Adhesive, Household Cement, Magic School Glue Gel, Paper Fix, Playtime Glue, Professional All Purpose Glue, Professional Wood Glue, School Glue Gel, Snif Proof Model Cement, Super Gel, Tack Tabs, Thick & Tacky Craft Glue, White Glue

Rubber Cement GluTube

Quick Grab

Quilters' GluTube

Satellite City: Hot Stuff, Special-T Hot Stuff, Super-T Hot Stuff

Sew-Art International: AppliHesive, Stix-A-Lot

Scotch: Heavy Duty Mounting Squares, Super Strength Adhesive

Signature: Crafter's Quick Relief

Solar-Kist: Easy Way Appliqué, Fine Fuse, Tuf-Fuse, Tuf-Fuse II, Tuf-Fuse III

Stanislaus Imports: Double Stick Adhesive, Glacé

Stiffy

TAC: Spray Stiff

Tandy: Crafts and Leather Cement, Magique

Tecnocraft: Hi&Lo, ONEDERGUN, Uni-Stik

Teflon

Testors: Cement for Wood Models

Therm O Web: HeatnBond Lite, HeatnBond UltraHold

Tip-Pen

Toner Plastics: Noodles

Top Flight Supreme

True Colors International: Stick & Hold, Stick & Hold for Fabric

Uchida: Marvy Glue Marker

Unstick

U.S. Star: Gloobies

Velcro: Velcro Glue-On Adhesive, Velcro Glue-On Tape, Velcro Sticky-Back Tape

Velvet Touch: Press 'N Wear

Warm Products: Steam-A-Seam

Washington Millinery: Bridal Glue

WD-40

Weldbond

Weldwood (DAP): Hobby & Craft Glue, Plastic Resin Glue

Wilhold: Plastic Resin Glue

W. H. Collins: Instant Vinyl, Unique Stitch

Zig 2 Way Changin' Glue

1

Stick with the Basics

- ▲ Glue Fundamentals
- ▲ Smart Product Choices
- ▲ Time-Saving Tools and Techniques

Glue crafting can be fast, fun, and simple, providing a wide range of charming projects for everyone from young to old, whether they're experienced or a beginner. But working with glues can be confusing.

By the time you read this book, there will be even more brand-new glues and adhesives on the market. **Why?** Because technology in this area is advancing rapidly and creating healthy competition in the fast-growing crafting market. **How can you keep up to date?** Read and follow the general guidelines in this chapter, then look for glues that will serve the specific purposes discussed in the remaining chapters.

Defining Glues and Adhesives

Originally the term "glue" referred to any adhesive material derived from an animal base, such as those made from hide, bone, or fish scales. Over the years, however, manufacturers have called many cements, adhesives, and other bonding agents "glues," so today crafting specialists use the terms "glue" and "adhesive" interchangeably. But even when the terminology is synonymous, the actual products are not all the same. For the best results, it's important to know which type of glue to use for every crafting project.

Why Is There So Much Confusion about Glue?

One of the questions most commonly asked of major glue companies and craft store personnel is, "What glue should I use to bond *X* to *Z*?" The next question usually involves exactly how to proceed. With all of the new products on the market in recent years and the wide range of choices, no wonder people are puzzled!

To top it off, the packaging and marketing of the glues themselves are sometimes confusing. In certain cases, the same product is marketed by more than one company (but if the brand name is the same, it should be the same product and you won't need both). More confusion results when one product is marketed under various labels from one company and then aimed at different end uses. To avoid purchasing the same product twice under different names, read the fine print and ingredients on the labels or ask a knowledgeable craft dealer.

No one glue works for every project, although some do work well for a variety of jobs. Some glues are specifically designed for only one type of craft

application. Some crafters feel most comfortable using a product developed for the exact type of craft they are working on, especially if they are working on a large job or doing a lot of work on the same type of project. Others prefer to use a multipurpose glue, testing first if necessary, to avoid having to make additional purchases for a variety of small jobs.

As you proceed through this book, you will learn which specific types of glues to use for which kinds of projects. Rather than analyzing the chemical composition of all the glues available to crafters (many of the ingredients are practically unpronounceable to the average person and the enormous assortment can be very confusing), this book separates glues by suggested end use. Now all you'll need to do for the best results is to decide what project you want to complete and select one of the possible choices listed.

Choosing the Right Glue for Your Project

No matter what materials you are gluing together, the adhesive you select should bond them well and last for a long time. After all, when you create your own special work of art, the last thing you want it to do is fall apart.

To help you determine which glue to use, I have divided the book into chapters by end use. Each chapter discusses the appropriate glues for each specific end use. The special applications are listed under each category of glue. For example, Chapter 2 discusses all the multipurpose glues you can use for porous or semiporous surfaces (those that contain open spaces that allow the glue to cling). Under White Glues in this chapter (one of the most common multipurpose glues), you'll learn how to use these glues for bread dough, découpage, papier mâché, and other special effects.

Chapters 3, 4, and 5 feature a wide variety of glues and adhesives developed and marketed for specific porous or semiporous materials, such as fabric and paper. In Chapter 6, you'll learn which powerful glues to use to bond hard, smooth, nonporous surfaces. Then Chapter 7 details a number of special-purpose glues not discussed in the previous chapters.

Hot-melt glues and their many crafting uses are included in Chapter 8. Glue guns have become increasingly popular with crafters because their glue can be substituted for many of the other glues (although not all), and they open up a number of interesting decorative possibilities.

In general, adhesives other than hot melts fall into two categories—water-based and solvent-based. (Fig. 1-1) The water-based glues are usually nonflammable, nonpolluting, and nontoxic, and they may or may not wash out (depending on the formula). The solvent-based glues have a faster drying time and are not water-soluble. They may be flammable or nonflammable, depending on the composition, and bond well to many surface types. In most chapters, you'll find a number of examples of both types.

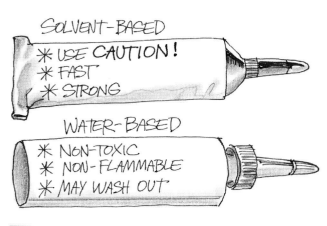

Fig. 1-1

Follow these five general guidelines for the best glue-crafting results:

1 **Consider the materials you are gluing together plus the end use and appearance of the project.**

▲ Will the glue used create a good bond between the specific elements being put together? There are many different formulations on the market for any type of job and some work better than others.

▲ Are the surfaces porous or nonporous? If a glue is fast drying, it may not be fully absorbed into the indentations on a porous surface and so it won't hold as well as it should.

▲ Will the surfaces need any advance preparation? Any surface should be clean, dry, and free of dirt and oil before gluing. Some surfaces, especially some woods (see Chapter 7), may need to be sanded as an additional preparation.

▲ Will the glue need to stretch on the finished project (such as an embellishment attached to a knit fabric) or will it remain stable? Some glues have been developed for stretchability. (Fig. 1-2)

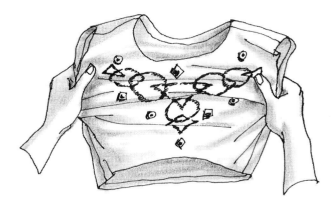

Fig. 1-2

▲ Does the glue dry hard or soft? For several end uses, a soft, flexible glue is important so that it can bend with the finished project.

▲ Do you want the project to be washable, dry-cleanable, or water-resistant? These factors are most important when crafting wearables and some home decor items or when you make a craft project that will be used outdoors.

▲ Does the glue dry clear, white, or colored? When the dried glue is visible on the finished project, you'll usually want it to be as inconspicuous as possible. The exceptions are colored craft glues or hot glue sticks that you will use decoratively.

▲ Will the glue give long-term durability and permanence if that is important on your project? (When the gluing job is temporary, this probably won't be a factor.) If the project will be subjected to excess stress, moisture, cold, or heat, the glue should remain as strong as possible under those conditions.

▲ Is the glue slow or fast drying? A fast setting glue works best for smaller jobs and for projects in which complicated alignment is not necessary, while a slow-setting glue will give you time to make adjustments and work on larger areas. "Setting" refers to the amount of time it takes the glue to harden. "Curing" time is how long it takes the glue to reach its maximum strength. When the glue is slow setting and/or slow curing, the elements being glued may need to be clamped or held in place by another means until the glue has completely bonded. (Fig. 1-3)

Fig. 1-3

2 **Among the best types of glues for your project, decide which is the easiest to use.** Everything being equal, which glue will be the easiest to apply? Will a fast-drying glue help hold parts of your project in place so that you don't have to clamp or otherwise secure them during the drying period? When a small dab of glue is all that you need, is it worth your time and effort to heat up a glue gun instead of just taking the cap off a tube of multipurpose adhesive? Will a glue gun be easier on a large job?

3 **Think about the cleanup required.** Can you wash away any excess glue with soap and warm water or will you need some type of solvent? If the product packaging doesn't tell you, read the general information given in this book, then test before using an unfamiliar glue.

Possible solvents include acetone-based nail polish remover, turpentine (mineral spirits), lighter fluid, and denatured alcohol, depending on the type of glue you are using. Some special solvents on the market have been formulated to dissolve only one type of glue, such as epoxy, super glue, or hot-melt glue.

4 **Always remember any important precautions.** Are you working with a toxic glue? This is of special concern for children, older people, those with health problems, pregnant women,

nursing mothers, and anyone who spends a great deal of time crafting with solvent-based glues. A hazardous substance can be accidentally inhaled, and some can be absorbed through the skin, often without our even being aware of it. Some substances may also be highly flammable. Read every label carefully for all the details. (Fig. 1-4)

Fig. 1-4

It's especially important for children to use nontoxic materials because often they don't understand or follow safety instructions (they might even decide to eat the glue). Look for products with a label that says "non-toxic" or "Conforms to ASTM D-4236" (which by law means that a toxicologist has reviewed and approved that specific brand's formula).

Another safety option is to call or write the manufacturer and request an "MSDS" (material safety data sheet) for the product you're using. This sheet will give you all the specific scientific details. Some chemically based products may be more harmful than others, although the current legal requirements for labeling do not allow the differentiation. "Chronic" ingredients can build up and cause irreversible damage, while "acute" ingredients can be harmful but will not have a cumulative effect. At this time, the only way you can be sure of the long-range effects of any product is to examine the MSDS and become more familiar with the product's ingredients.

Most healthy adults won't be significantly affected by a minimal use of the chemicals in craft glues now on the market. However, always diligently follow the instructions printed on the label in order to avoid any unnecessary hazards (both immediate and long-term). Even common rubber cement can be harmful if used incorrectly—here's a precise list of precautions from the jar of a well-known brand:

"CONTAINS PETROLEUM DISTILLATES. HARMFUL OR FATAL IF SWALLOWED. KEEP OUT OF REACH OF CHILDREN. Use in a well-ventilated area, away from open flames, sparks, pilot lights, cigarettes and other heat sources. Avoid frequent or prolonged skin contact. If it gets into eyes, flush immediately with water. Call physician. If swallowed, do not induce vomiting. Call physician immediately."

When your project calls for a glue or adhesive that is potentially harmful, follow the instructions on the label carefully. Throughout the book, I've included these icons to remind you to read the label first and proceed with care.

5 **Evaluate the price in relation to the job.** You don't need a bazooka to kill an ant. Likewise, you don't need to go out and buy the most expensive high-tech glue on the market when a white craft glue will do the job. Especially when you are working on a large project or when you do lots of crafting, you can save considerable time, effort, and money by making smart choices, just as long as the less expensive glue does the job you want it to do.

Using the Proper Tools and Accessories

You won't need to run out and buy expensive or elaborate equipment for most gluing projects—many common household items work beautifully. But having the right type of supplies readily at hand can make your work much easier:

▲ **Work surface protectors**—My favorite is a roll of waxed paper because it's inexpensive, doesn't take up much space to store, the exact amount you need can be torn off easily, and glue won't soak through it. (Fig. 1-5) Other good work surface possibilities include coated freezer wrap, plastic wrap (it sticks easily to a damp counter), aluminum foil, and plain disposable paper (newsprint may smear). One good option is a large desk pad with tear-off

Fig. 1-5

sheets—you can leave the pad out all the time, sketch designs or make notes on it, and then throw away the top sheet when you're finished with the project.

▲ **Utility trays and containers**—Save plastic (or aluminum) food trays and containers. Use them for mixing, organizing project components, and holding small embellishments (such as the beads or glitter discussed in Chapter 4).

▲ **Glue applicators**—Use toothpicks, bamboo skewers, wooden craft sticks, plastic eating utensils, or collar stays to help spread and position glue. Keep handy several brushes in a variety of sizes. Many glues wash out with soap and water, but if you're working with one that doesn't, you might want to consider using a disposable sponge brush for easy cleanup. To precisely position a drop of glue for bead work or another exacting job, a *Magic Glue Wand* with a round angled tip, available from Clotilde (see Resources), works beautifully.

▲ **Fine-line applicator tips and bottles**—To apply glue in a very fine line or with pinpoint accuracy, try one of the fine-line bottles or tips designed for fabric paint. These work best with the thinner, more liquid glues. The widely available *Tip-Pen* set (now marketed by Plaid) has metal tips in four sizes and extension caps to fit most popular brands. (Fig. 1-6)

To make a small temporary tip on the neck of any opened glue bottle, wrap a piece of cellophane tape tightly around it diagonally. Manipulate the tape for the exact-size opening you want and leave the end unwrapped for easy removal. (Fig. 1-7)

▲ **Applicator syringes**—When you're doing a big job or doing a lot of gluing with the same glue, try a syringe for easy application. One popular model is Aleene's *Fine Line Syringe*. (Fig. 1-8)

▲ **Disposable gloves**—As a precaution with toxic materials (and because of my manicure), I keep a box of lightweight disposable latex gloves (available at any drugstore) on my work table. For only pennies, you can quickly slip on a pair, keep your hands clean while your fingers still have complete mobility, then peel them off and throw them away when you're done. I highly recommend disposable gloves for anyone who works with glues often, especially solvent-based glues.

▲ **Hand lotion**—One new product from Signature Marketing & Manufacturing, Inc., is *Crafter's Quick Relief*, another excellent item to keep on hand. It moisturizes and smooths your hands if you must wash them often after working with glue. In addition, it has special ingredients to help relieve cramps and stiffness after doing a lot of detail work.

▲ **Cleanup supplies**—Keep a box of pop-up facial tissues handy to clean up any small

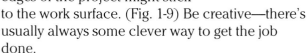

Fig. 1-6

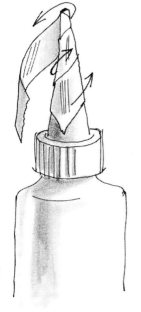

Fig. 1-7

messes quickly or to remove glue from tube or bottle tips before resealing them. When working with water-based glues, a damp rag or a container of baby wipes can be helpful. Paper towels are another craft-table staple.

▲ **Drying aids**—Clamp a project together for drying with clothes pins, long smooth bobby pins, rubber bands, or removable tape. Also consider using the wider paper clamps or the giant clamps designed to close snack bags.

I use the plastic center supports from delivered pizza as small drying towers when the edges of the project might stick to the work surface. (Fig. 1-9) Be creative—there's usually always some clever way to get the job done.

▲ **Other helpful tools and notions**—I find these items indispensable on my work table: pliers (both regular and needle-nose), wire cutters, tweezers, craft scissors, a ruler, a craft knife, tape (cellophane and masking), pins (including corsage pins and T-pins), pencils and marking pens (including permanent, air-erasable, and water-soluble), clear ace-

Fig. 1-8

Fig. 1-9

tone-based nail polish remover (as a solvent for some glues), Magic American *Goo Gone* (it cleans up a variety of glues), a nail (for opening some tubes), and a small pointed-tip bottle filled with water (for quick cleanups or for slightly diluting some water-based glues).

▲ **Craft supplies for specific projects**—The other types of materials you keep on hand will depend largely on what crafts you enjoy doing. Beginning with Chapter 2, the instructions for specific projects will include all the supplies you'll need for each one. For example, your project may call for "dimensional" fabric paint which will remain raised or puffed on the surface after application (other fabric paints flatten into the fabric or spread out over the surface).

General Gluing Tips

As you read through this book, you'll find special uses and application how-tos for a wide variety of crafting glues. Some general guidelines, however, will apply to all of your gluing projects:

1 Read and follow the instructions for the type and brand of glue you are using, before beginning your project. (Fig. 1-10) Adhesives

Fig. 1-10

vary widely, even when they accomplish similar end results. Therefore, the correct way to apply them often differs as well.

It's safe to say that most common gluing problems stem from not following the manufacturer's directions. Read the label or accompanying literature, especially when using a glue for the first time—you'll usually find a great deal of helpful information.

For example, while some glues are applied in a single coating, others bond more strongly when applied to both surfaces, allowed to dry for a short time, and then the two pieces are brought together and exactly positioned. This is sometimes referred to as the "contact" method because the pieces being glued will stay where they are initially positioned when the contact is made and won't slide around for further alignment. The contact method is commonly used to apply cements, although some newer crafting cements recommend it only for porous surfaces where gaps need to be filled.

The manufacturer's instructions will usually describe the exact application procedure (including the suggested drying time), what materials the product works best for, and any precautions you should observe. Don't risk losing valuable time and materials by gluing first and reading the how-tos later!

2 When in doubt, always test first. Let's say you want to glue fabric to plastic foam on a small project, and you have on hand a tacky glue, household cement, and *Amazing Goop.* Can you use one of these or should you go out and buy a glue made specifically for plastic foam? Do a quick test with scraps of both materials and check the results. In this case, both the cement and *Amazing Goop* will probably dissolve the foam, but the tacky glue should work beautifully. (In fact, it may soak through the fabric less than a glue designed specifically for plastic foam.)

Testing is an essential first step on many new gluing projects. It can be done quickly and easily, so why flirt with disaster instead?

3 Be sure to open new tubes or bottles of glue correctly. To prevent the glue from evaporating or leaking before it is sold, many tubes and bottles come sealed. You'll usually need to pierce the metal opening of a tube, remove a plug, cut off the tip of a plastic applicator, or screw open the bottle. (Fig. 1-11) When all else fails, don't just squeeze as hard as you can—read the directions instead.

REMOVE & CLEAN APPLICATOR TIP

RESEAL WITH ORIGINAL CAP

Fig. 1-12

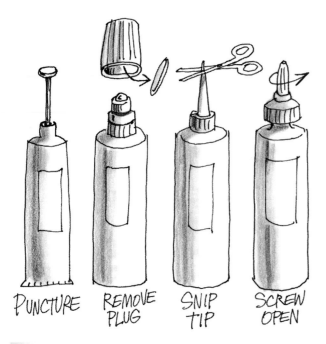

PUNCTURE   REMOVE PLUG   SNIP TIP   SCREW OPEN

Fig. 1-11

"Once I presumed the glue was very (very) slow in coming out of the bottle. If I squeezed harder, out it would come. It came out all right—the side of the bottle split. What a mess! Always check to see if the plug or seal has been removed when the glue isn't coming out the tip."

Nancy Ward, *Fabric Painting Made Easy*

How to open the container may seem obvious, but after using some glues, you will find that the tip can become clogged as the excess dries. It's second nature for many of us to think that exerting a little extra force will solve this problem by releasing the clog. But beware! Even a clogged container can burst. Use a large pin, a bent paper clip, or a section of wire to clear the opening. If the applicator tip is removable, take it off and work gently from either direction. (Fig. 1-12)

When opening a metal tube that doesn't have a special applicator tip to be screwed on later, use a pin to make only a tiny hole—the glue will come out more slowly and you'll waste less.

4 **Glue only clean, dry surfaces.** Before applying glue to a project, be sure the area to be glued is free from moisture, dirt, dust, and grease.

5 **Keep your work area, tools, and hands clean and glue-free.** Using disposable paper on the work surface and wearing disposable gloves are a big help. Then use facial tissues to wipe off the tip of a bottle or tube after you've finished using it and before putting the cap back on. You'll be happy you did because the next time you use it, it won't be sealed closed. (Fig. 1-13)

Also wipe off tweezers, pliers, scissors, or other tools before any glue dries on them. By doing this, you will have very little cleanup with most projects. Dried glue is usually much harder to remove.

6 **Work patiently.** Don't rush when working with glues. A neat, careful application will look professional, while rushing will probably produce a sloppy job. Even if the glue is very fast drying (such as a super glue), plan each project step carefully and work methodically for the best results.

Fig. 1-13

7 **If necessary, use more than one glue in the same project.** One project may need different glues for various parts or you may need one glue to give a quick temporary hold while another hardens overnight. Think through any project carefully before you begin.

8 **Use special techniques with some water-soluble glues.** Because some water-soluble glues can separate slightly while sitting on the shelf, be sure to shake the glue well before using it. Then, holding the bottle upside down, snap your wrist firmly two or three times to work the glue down into the tip and remove any air bubbles before you begin to apply it.

9 **Beware of old glue.** Don't use a glue you've had on hand for a while without testing it first. All glue products have a limited shelf life, so test their effectiveness after one year. You certainly don't want to spend time and money on a terrific project only to have the glue fail you!

10 **Store any opened glue properly.** Prolong the life of an opened glue by resealing the container tightly. If the cap was removed and an applicator tip put on, replace the cap. Clean out the tip as much as possible and then soak it in the solvent recommended for that type of glue.

If you'll be using the glue again soon, stand tubes with long applicator tips on end so the glue can drain back down into the tube and won't dry out and clog the tip. An old mug or small jar makes a good stand for holding the tube upright. Slide a stainless-steel corsage pin down the tip to seal the opening temporarily. (Fig. 1-14)

For the avid crafter, there are stands on the market to hold some glues upside down—the glue will settle at the tip end and will be immediately ready to come out at the first gentle squeeze. Make your own ready-glue stand by turning the capped bottle upside down in a mug or jar. Or store the closed glue containers upside down with the tips embedded in a block of plastic foam.

11 **Let a washable glue dry thoroughly before laundering it.** By washable, I mean a permanent glue that has been developed to withstand washings, not one that washes out. Some experts say that one week (or even two) is a minimum drying time before the first washing. Keep in mind that even when a glue appears dry, it may not be.

12 **Keep up to date on your favorite crafting areas.** Whether you love woodcrafting, wearable art, amateur jewelry making, or any other craft, you'll usually find stores, mail-order catalogs, magazines, and helpful books on the subject. See Resources at the back of the book for some good ideas on where to begin.

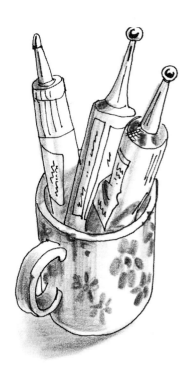

Fig. 1-14

2

The Workhorses of the Crafting World

- ▲ White Glues
- ▲ Tacky Glues
- ▲ Clear Craft Glues
- ▲ Super Glue Gels
- ▲ High-Tech Adhesives

Time and time again, for bonding porous or semi-porous materials you'll reach for a multipurpose craft glue. The wide range of these versatile multipurpose glues are formulated for use with a variety of materials, including fabric, paper, soft wood, dried or silk flowers, feathers, felt, and ribbon.

Other multipurpose glues have been developed to form a strong bond on smooth, nonporous surfaces and are discussed in detail in Chapter 6. To confuse the issue, some of the newer high-tech glues work well for both types of surfaces.

Many of the multipurpose craft glues discussed in this chapter on bonding porous and semiporous materials are water-soluble and nontoxic. They will provide an adequately strong bond for a variety of projects. These glues are quick and easy to use and pose no health danger. For safety, water-soluble nontoxic glues should always be used when children are doing crafting projects.

When you need a stronger, longer-lasting bond for your general crafting projects, another type of glue may be necessary, such as a super glue gel (page 19) or a high-tech adhesive (page 22). These glues are often more expensive, may be harder to use and clean up, and can pose fire and health dangers if used incorrectly. But they do work well in some places where a water-soluble craft glue will not. Always read the label and follow the directions carefully.

WHITE GLUES

Some white craft glues, such as Elmer's *Glue-All* and Delta *Sobo,* have been available for years and are handy in many situations. Most white glues are nontoxic and wash out with soap and warm water before hardening. Those products specifically labeled "school" glues (by Elmer's, LePage's, Aleene's, and Ross, for example) and a few other white glues are formulated to wash out even after they have hardened completely. Presoaking is sometimes necessary.

Formulas used for white glues vary. When you hear a glue referred to as a PVA (or polyvinyl acetate), it falls into this category. Because of their high water content, most white glues don't dry quickly and don't form as strong a bond as some other glues, so you may want to use them mainly for lightweight projects. None of the white glues dry completely flexible and many are quite stiff when hardened, but this usually isn't a problem when you use them for a nonflexible project.

Some white glues are less liquid and runny and some dry faster than others. Certain formulas withstand multiple freeze-thaw cycles and others do not—read the label. And although most white glues dry clear, some are clearer than others. After experimenting, you'll likely find your favorite brand or brands for the specific types of projects you enjoy.

Use white glues mainly for bonding porous and semiporous materials. Although some brands say that they work for ceramics, metal, and glass, they aren't usually as strong as other glues on the market. If you do want to use a white glue for these heavier or less porous materials, always test first.

Brand names of white glues, in addition to those already mentioned, include: Aleene's *White Glue,* Bond *Stik,* Crayola *Art & Craft Glue,* DAP *Weldwood Hobby & Craft Glue,* Franklin *Home, Shop & Craft Glue,* Kony Bond *Multi-Purpose White Glue* (no-run formula), LePage's White Glue, Loctite *Desk Set Safe & Easy White Glue,* and Ross *Professional All Purpose Glue* and *White Glue.*

Use any of these glues for a wide variety of fun kids' projects. One easy example is to incorporate materials on hand to create cute "Fanciful Friends." Designed by Karen Dillon of Walnut

Fig. 2-1

Creek, California, they're great for decorating gift packages, cakes, plants, and more. Quickly glue decorative tidbits, including sequins, jewels, beads, yarn, ribbon, bits of fur or pet hair, and glitter to assorted plastic utensils. Use a permanent marker to add detail. (Fig. 2-1)

White PVA glues also come in colored form. Look for Elmer's *GluColors,* LePage's *Hot Tropicals,* and Ross *Playtime Glue.* (For the easiest use by children, store them with the tips upside down (see page 9). In addition to colored glues, Elmer's recently introduced glitter colors and glues that dry black or solid white to complement their color line. *Crayola* also has glitter glues in pen form.

Bread Dough

Crafters of all ages will enjoy the simple art of no-bake bread dough projects. All it takes is white bread, white glue, and a little time to mix and shape the dough. Create colored dough by using colored glues (see above) or mix a white glue and color it with food coloring, acrylic paint, or nondimensional fabric paint. (For children, use either food coloring or colored nontoxic glue.)

You can also use tacky glues (see page 17) for bread dough art. Aleene's recommends their original *"Tacky" Glue* for a ceramic look and their *Thin Bodied Tacky Glue* for a porcelain finish. Their *White Glue* is not recommended for bread dough crafts.

One basic bread-dough formula is one slice of bread to one tablespoon of white or colored glue:

1. Remove the crusts from the number of slices you'll need for your project and tear the bread into small pieces (or use a food processor). Put the bread in a mixing bowl or in a heavyweight zip-lock baggy. (I prefer using a baggy to eliminate cleanup. Just throw away the baggy when done instead of having to wash out the bowl before the glue dries.)

2. Pour glue over the bread in a ratio of 1 tablespoon to 1 slice of bread. Knead the dough for 8 to 10 minutes or until it becomes smooth and workable. At first the dough will be sticky and difficult to mix, but keep kneading. (Fig. 2-2)

    Because the moisture in the bread or the consistency of the glue (and even the weather) can cause variations in the dough, you may need to adjust the mixture. After working with the dough for at least 8 to 10

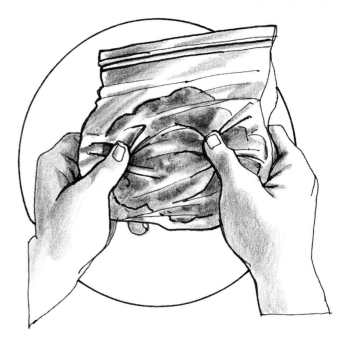

Fig. 2-2

minutes, add flour or more crumbled bread if it's too sticky or add a little more glue or a few drops of water if it's too stiff.

3. If you didn't use colored glue to make your dough, add paint or food coloring after the dough is smooth. Color only the amount of dough you'll need for each color in your project. Keep the rest in a sealed zip-lock bag in the refrigerator and it will last for months. (Keep any leftover colored dough in separate baggies.)

    To add color neatly, make an indentation in the center of the glue, add only a few drops of paint or food coloring (you can always add more later if necessary), fold the glue over the drops, and knead until it's well mixed. You may have to mix colors to achieve the exact look you want.

4. Roll out the dough on a smooth, clean surface as you would a cookie dough (dusting with flour if necessary) and cut or stamp shapes from it. Or use your fingers to mold small pieces of unrolled dough into petals, leaves, and other shapes for your project. The dough will often stick to a rough or porous surface when pressed on while moist, or you can let the dough dry and glue it on later. Some projects will take only hours to dry, while other, larger pieces can take several days.

5 Leaving the dried dough projects unsealed will give them a matte finish, or you may choose to coat them with a clear acrylic sealer or paint to add a sheen and give them more durability.

PROJECT: Sequinned Dough Ornaments

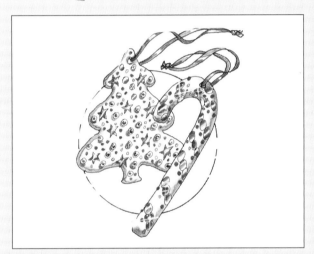

Make these expensive-looking ornaments to trim a tree or to give as a special holiday gift to a friend. (Fig. 2-3)

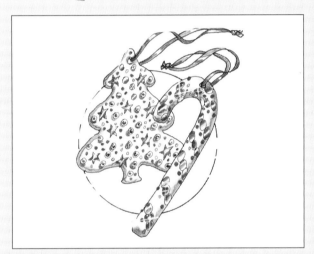

Fig. 2-3

MATERIALS NEEDED

· 8 slices of day-old white bread (makes several ornaments, depending on the size.

· 8 tablespoons of white glue

· A large zip-lock baggy or a mixing bowl.

· Food coloring or acrylic paint

· Small cookie cutters in tree, candy cane, or other holiday shapes

· Flat sequin jewels (from a craft store)

· Clear acrylic sealer

· Narrow ribbon or cord

HOW TOS

1. Make the bread dough, following steps 1 through 4 (page 13), and roll it out to about ⅛″ thickness.

2. Using the cookie cutters, stamp out the ornament shapes. Form a hole for hanging in the top of each ornament—the end of a plastic drinking straw works well for this.

3. Let the ornaments dry, turning them occasionally so that both sides dry evenly.

4. Using a skewer or wire coat hanger, suspend each ornament from the top hole. (Fig. 2-4) Completely coat the ornament with acrylic sealer or paint and position the sequin jewels while wet. After drying, check to see that all sequins are firmly attached. If not, reapply them with a drop of white glue.

5. Add one or two more coats of sealer over the sequins and the entire surface. When dry, tie a short length of ribbon or cord through the hole of each ornament.

SKEWER OR WIRE THROUGH ORNAMENT HOLE.

BOXES, BRICKS, OR OTHER SUPPORTS HIGHER THAN ORNAMENT.

Fig. 2-4

Découpage

White glue has been a traditional favorite for découpage, a craft in which paper cutouts are glued to wood, metal, or another stable base and the entire surface is coated with a sealant. (Fig. 2-5)

Fig. 2-5

Apply slightly thinned white glue either to the back of the cutouts (usually favored) or to the base surface. After the glue just begins to set, position the cutouts. A number of sealants can be used for découpage, including lacquer, shellac, and polyurethane. Another option is to use a combination primer, adhesive, and sealant, such as Plaid *Mod Podge*, which comes in either a matte or gloss finish. Apply several coats and finish with a topcoat of clear acrylic sealer to prevent tackiness.

White glue can also be used as a découpage coating, then apply a topcoat of clear acrylic sealer. In most cases you'll need to dilute the glue with water for a brushable consistency. Colored glue is another topcoat option, giving a tinted effect.

A newer product on the market is Plaid's all-in-one découpage treatment, *Royal Coat*, which glues and seals in only two to three coats and does not require a sealer coat. Another product, Beacon *Liquid Laminate* (see page 52), produces a similar effect.

Papier Mâché

White glue is often used as a binding ingredient in the age-old craft of papier mâché. Used full strength, the glue dries rapidly and gives the project a tough surface. When diluted (usually 3 parts glue to 1 part water), it seals and fills well, creating a smooth, paintable surface. Test first before beginning a large project.

For specific instructions for making papier mâché, see page 56, substituting white glue (or diluted white glue) for wheat or "wallpaper" paste.

Special Effects

Both white and tacky glues can be used to produce a variety of interesting finishes on your craft projects.

Glazing. You've just read about how to dilute a white glue with water and use it as a coating for découpage. This technique also can be used for a tinted brush-on glaze or coating over any printed or textured surface. Either dilute one of the colored glues on the market or add acrylic paint to a diluted white glue. For a look more like enamel or ceramic, brush on several coats of glaze. Depending on the end use, you may need to seal the project as you would for découpage in order to prevent tackiness and add durability.

Marbelizing. Produce a marbled effect by swirling a second colored glue through a glazed surface (see above) while the glue is still wet and before applying a sealer. To swirl the second color, use a toothpick, cotton swab, or anything else that gives you the desired effect. (Fig. 2-6)

Fig. 2-6

Enameling. Create a heavier finish by using the colored glue at full strength instead of diluting it for a glaze. Cover the project surface completely with a smooth, thick layer of glue. Then while it is still wet, add drops of one or more additional colors, using a toothpick or pin to perfect the design. After the surface dries, coat it with a gloss-finish sealant such as *Right-On* from Aleene's, the company that perfected this enameling technique. (Fig. 2-7)

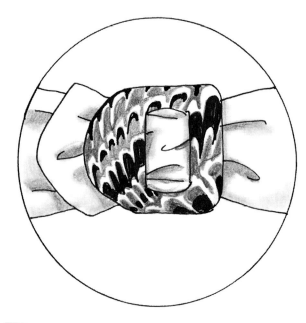

Fig. 2-7

Texturizing or embossing. Add texture to any project by creating raised designs with drops or lines of glue. (Fig. 2-8) For an accurate design, use a syringe or fine-line applicator (see page 6) or dot the glue on with a toothpick. Let the glue dry thoroughly before glazing or painting over it.

Fig. 2-8

Salt glazing. Coat the project surface with colored glue. Then, while still wet, cover the glue thoroughly with table salt. Rub off the excess when dry to leave a pretty sugar-coated appearance. (Fig. 2-9) Use more than one color of glue to increase your design options. (See Chapter 4 for many more embellishment techniques.)

Fig. 2-9

TACKY GLUES

Often classified as a type of white glue, tacky glue is specially formulated for a quick "tack" or bond. These glues usually dry clear and are flexible and nontoxic, but they can wash out.

Tacky glues often can be substituted for the white glues in the applications previously discussed. For some projects, a slower tack is helpful so that you have more time to move and adjust the items being glued. For others, the quicker bond of a tacky glue is needed to hold components easily in place until fully dry. The quickness of the tack sometimes varies with the consistency of the glue:

▲ **Heavyweight tacky glues** have been developed to work somewhat like a hot glue gun with a very quick tack. When positioning light items, such as in a dried floral arrangement, these glues quickly hold the components in place at any angle until they are fully set. (Fig. 2-10) Heavyweight tacky glue also works well for any other crafting projects that are difficult to hold in an exact position while the glue dries. Brand names of heavyweight tacky glues include: Accent *Tacky Glue,* Activa *Mighty Tacky,* Aleene's

Thick Designer Tacky Glue, Bond *484:Tacky,* and Delta *Velverette.*

▲ **Regular tacky glues** are useful for a wide range of crafting projects. They have a medium consistency, are easy to apply, and they clean up with water. Although they usually don't have quite as quick a tack as the heavyweights, they can be applied more smoothly. Brand names of regular tacky glues include: Aleene's original *"Tacky" Glue,* Bond *Instant Grrrip,* Bond *Squeezable Tacky,* Borden *Craft Bond I,* Delta *Quik 'n Tacky,* Jurgen *Tacky Craft Glue,* Plaid *Tacky Glue,* and Ross *Thick & Tacky Craft Glue.*

▲ **Lightweight tacky glue** is available from Aleene's, marketed as *Thin Bodied Tacky Glue.* The glue is chemically thinned to go on smoothly, in a light layer, and is also promoted for making "porcelain-look" bread dough. Using the recipe on page 13, this thinner glue will make a smooth dough that looks similar to porcelain, while bread dough made with a white or regular tacky glue will have more of a ceramic appearance.

Even though some tacky glues are "water resistant," they are not recommended for use on clothing or other projects that you'll be washing regularly. Most tacky glues have been developed to prevent soaking through fabric easily and showing on the right side. If you do have trouble with soak-through, Aleene's suggests applying tacky glue to the fabric and letting it set for five minutes before attaching it to the project. Then wait another five minutes before firmly pressing it in place.

Formulas and characteristics of tacky glues also vary from brand to brand. For example, a thicker glue is not always the tackiest. Always remember to test first to see how any glue will work for the specific job you have in mind.

Fig. 2-10

PROJECT: Bits-and-Pieces Necklaces

Designed by Jolly Michel of Stitzer, Wisconsin, these one-of-a-kind necklaces feature odds and ends of fabric, ribbon, buttons, feathers, and charms. (Fig. 2-11)

Fig. 2-11

MATERIALS NEEDED

· One 1-½" by 45" strip of print or solid color fabric, cut on the straight of grain (or, if you don't sew, substitute a 45" length of flexible ¼"-wide ribbon; then skip step 1)

· Woven ribbons in various widths to coordinate with the fabric strip

· Assorted buttons, beads, charms, jewels, feathers, and any other appropriate embellishments

· Metallic and fine feathery yarns and threads

· ¾" bias tape maker (if you're doing step 1)

· Tacky glue, such as Bond *Instant Grrrip*

· Optional: wire cutters for button-shank removal

HOW TOS

1. Feed the fabric strip through the bias tape maker to create a strip ⅜" by 45". Top-stitch to close the open long edge.

2. Fold the ends of the strip to the center, gluing them in place on the underside. (Fig. 2-12) To create a closure, sew a small button on one end and use the tacky glue to adhere a short length of ⅛"-wide ribbon to the other end, forming a loop just large enough to fit smoothly over the button. Hide the loop ends between the layers.

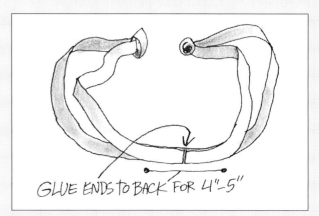

GLUE ENDS TO BACK FOR 4"-5"

Fig. 2-12

3. Tape the necklace base to your work surface in a V shape. Working from the center out, decorate 4" to 5" at the front by gluing on layers of ribbon loops, buttons, and other decorative items. Use an odd number of buttons for visual interest. On buttons with visible holes, thread through decorative yarn or thread before attaching. For buttons with a pronounced shank, use wire cutters to remove the shank before gluing.

4. Twist and glue more decorative yarn or thread around the buttons as filler and tie any additional charms or beads to the yarn or ribbon ends for a finishing touch.

CLEAR CRAFT GLUES

For lighter-weight projects, you'll find a clear multi-purpose glue helpful to keep on hand. Most glues in this category (usually a polyvinyl alcohol formula) are nontoxic and wash out more easily than white or even white school glue, making cleanup simple.

Although they are susceptible to moisture, clear craft glues work well for paper, cloth, and other lightweight arts and crafts. Brand names include: Best-Test *Grade School Glue,* Itoya *O'Glue,* Kony Bond *Easy Paper Glue,* LePage's *Arts & Crafts Glue* and *Clear Glue,* Pentel *Roll'n Glue,* and Eberhard Faber *UHU Liquid Glue Pen.* For ease of application, some clear glues are marketed in pen form with a sponge or roller applicator.

Multipurpose glue gels are formulated differently from the more liquid clear glues to prevent runs and drips. They also dry more slowly to allow any necessary repositioning. Glue gels may appear to have a blue or purple tint in the bottle, but they dry clear. Brand names include: Elmer's *School Glue Gel,* Loctite *Desk Set Clear Gel Adhesive,* Ross *Magic School Glue Gel,* and Ross *School Glue Gel.*

SUPER GLUE GELS

Super glue (chemically a cyanoacrylate) is discussed in Chapter 6 as a power bonding agent. This liquid super glue is an instant adhesive, but it works much better on smooth surfaces that fit together tightly. For this reason, a gel form of super glue was developed specifically for porous and semiporous materials or for surfaces that don't fit snugly together. The super glue gel won't run and will fill in a porous or uneven surface area.

The biggest advantage of any super glue is that it forms a powerful bond in seconds (although the bond will not be as durable as many of the high-tech adhesives discussed next). On the flip side, the quick bond does not allow for repositioning if the components are not aligned perfectly.

Use super glue gel for embellishing clothes and accessories with jewels and rhinestones, wet reed projects, and any other small quick crafting jobs. Super glue gel is often not recommended for glass, plastic foam, rubber, or some plastics—always test first.

Super glue gel is water-resistant, but it is not the best choice for washable projects (see Chapter 3). Although super glue gel is relatively expensive, you'll need only a small drop per square inch. Using more glue will actually weaken the bond.

Widely distributed name brands of super glue gel include: Borden *Krazy Glue Gel,* Devcon *Super Glue Gel,* Duro *Quick Gel,* Kony Bond *Super Glue Gel,* Loctite *Quicktite* (in a new, easy-to-use *Gelmatic* dispenser), Loctite *Quicktite Super Glue Gel,* 3M *Super Glue Gel,* and Ross *Super Gel.*

PROJECT: Shopping-Bag Refrigerator Magnet

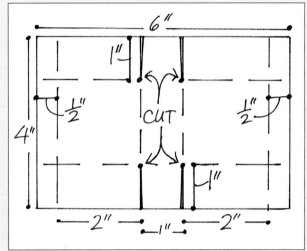

Hate to waste those last scraps of a pretty wrapping paper? Now you can turn them into little refrigerator ornaments to suit any season or occasion. (Fig. 2-13)

Fig. 2-13

MATERIALS NEEDED FOR EACH MAGNET

· One 4" by 6" rectangle of wrapping paper to suit the season (Christmas, Easter, Halloween) or for a special event (birthday, new baby, wedding)

· 10" of 1/8"-wide ribbon (twine or cording can be substituted)

· One small magnet (available in craft stores and catalogs)

· One 6" square of tissue paper in a coordinating color

· Clear craft glue

HOW TOS

1. Crease and cut the paper rectangle following the pattern. (Fig. 2-14)

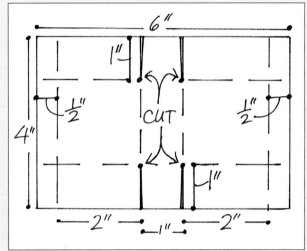

Fig. 2-14

2. Fold and glue the two 1/2" hems on the rectangle ends.

3. Cut the ribbon in two equal lengths and glue the ends on top of the folded hem sections, approximately 1/2" from the center of each end.

4. Fold and glue the bag sides together with the 1" center cutouts glued upward. (Fig. 2-15)

5. Glue the magnet at the center back of the bag.

6. Grasp the tissue square in the center and stuff it firmly down into the bag.

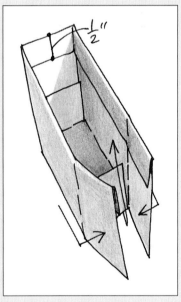

Fig. 2-15

PROJECT: Tablecloth Tamers

Hold down a flyaway tablecloth with decorative weights attached to the corners. Super glue gel grabs quickly on any porous or semiporous surface to make this project in a jiffy. (Fig. 2-16)

HOW TOS

1. When using shells which need to be weighted, add a drop of super glue gel to the weights and hold them in position in an inconspicuous place on the shells until dry—at least 15 seconds. (Fig. 2-17) Remember to use only about one drop per square inch for the strongest bond.

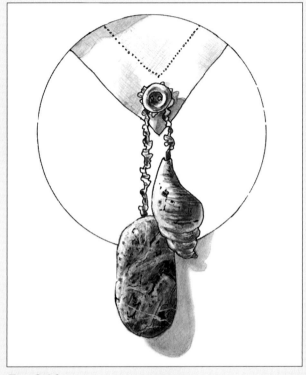

Fig. 2-16

Fig. 2-17

2. Cut the ribbon into four equal lengths and glue the end of each to a weighted shell, stone, or other heavy ornament.
3. Sew a button at each corner of the tablecloth and loop the center of a ribbon completely around each button, letting the weighted ends hang down to anchor the cloth.

MATERIALS NEEDED

· Eight shells, attractive stones, or other weighty items
· Fishing or curtain weights that fit inside the shells (for heavier items, weights are unnecessary)
· 1-⅛ yard of ¼"-wide grosgrain or crinkled ribbon
· Four large buttons
· Super glue gel, such as Loctite Quicktite

HIGH-TECH ADHESIVES

Recently, many new high-tech adhesives have been developed to make a wide range of gluing jobs much easier. Even though a product may not have been originally intended for crafting purposes, we can now reap the benefits.

Most high-tech adhesives are solvent-based and therefore are not water-soluble. One notable exception is *Weldbond* from Frank T. Ross & Sons. This versatile product can be used full strength or diluted with water. In addition to a wide variety of construction uses, it bonds almost anything to anything else, including both porous and nonporous materials, wood, glass, and ceramics. *Weldbond* is not recommended for some types of plastic, rubber, or cast metals or for any container that will be used to hold hot liquid.

Weldbond is nontoxic and odorless; it cleans up easily with water, sets completely within an hour, and is transparent when dry. For a free ten-page guide on the product, send a self-addressed postage-paid envelope (U.S. postage for U.S. addresses) to: Frank T. Ross & Sons, 6550 Lawrence Ave. E., Scarborough, Ontario, Canada M1C-4A7.

Other high-tech adhesives vary widely in formula, but all have in common the ability to bond strongly a wide range of materials. They are also quick bonding (usually in only a few minutes), fast drying (most set completely in 24 hours), and fill in gaps well. They are also clear and flexible when dry, most with good water resistance.

Common solvent-based brand names include: Bond *Victory 1991,* Eclectic E-6000 and *Amazing Goop Adhesive Products* (including *Crafter's Goop, Amazing Goop,* and *Household Goop*), Loctite *Stik 'n Seal,* Pacer *Zap A-Dap-A Goo, Quick Grab,* and Eberhard Faber *UHU Bond-All.* All of the Eclectic products are made with a similar formula but vary in viscosity.

Read each label carefully for specific use instructions—they do vary. Most will dissolve plastic foam, so be sure to test when in doubt.

Recommended solvents or cleanup procedures are included on each label and depend on the specific formulation of the product. For example, after using an applicator nozzle with *Amazing Goop* products, unscrew the nozzle, clean it out as much as possible, and soak it in lacquer thinner or xylene (from a home center or hardware store) in a well-ventilated area. Replace the original cap for storage. To ensure easy cap removal for the next usage, apply a light coat of petroleum jelly to the threads of the tube before replacing the cap.

Many high-tech adhesives can be somewhat difficult to squeeze out of the tube. Use a tube gripper to make the job much easier. The gripper delivers precise amounts and allows you to get the maximum adhesive out of every tube. One good gripper on the market is the Amazing Goop Adhesives *Tube Gripper.*

PROJECT: Fast Floral-Topped Box

Glue a small dried floral arrangement and ribbons to the top of any size straw box to create a unique decorator item. Use the box for potpourri, assorted trinkets, or to package a special gift. (Fig. 2-19)

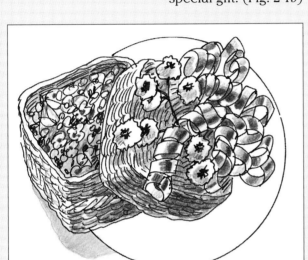

Fig. 2-19

MATERIALS NEEDED

· One straw box with a fitted lid
· Dried floral sprigs of a size and quantity to fit the box lid
· Metallic curling ribbon (or any other coordinating ribbon tied into an interesting bow)
· High-tech adhesive, such as Crafter's Goop

HOW TOS

1. Arrange the dried floral pieces and glue them in place on the box lid, following the instructions on the glue label.
2. Curl the ribbon (or tie a bow) and glue it on top of the floral stems.

3

Speedy Stitchless Sewing

▲ Fabric Glues
▲ Fusibles
▲ Repositionable Appliqué Adhesives
▲ Thread and Fiber Sealants

Attach trim, apply an appliqué or emblem, and secure a hem quickly and easily, all without using a needle and thread. Put the glues and fusibles in this chapter to work effectively on a wide range of stitchless sewing, decorating, and crafting projects.

FABRIC GLUES

Chapter 2 featured a variety of multipurpose glues that work well on cloth, but many additional glues have been developed and marketed specifically for no-sew fabric projects. White and tacky glues work well but wash out when a project is laundered. Aleene's *No-Sew* is a fabric glue that also washes out, but it was developed to hold through several dry-cleaning cycles. A few other fabric glues are washable and also dry-cleanable (see below). **Note:** Because dry-cleaning solvents vary, always test a sample before dry-cleaning your finished project.

Most fabric glues were formulated to glue fabric to fabric or to glue trim, lace, and leather to fabric. Many of the glues listed here also work well for gluing other types of embellishment materials to fabric and often can be used as effectively as the embellishment glues listed in Chapter 4. In addition, a number of the glues marketed as embellishment glues work well as fabric glues. Check the label.

Glue formulas and fabric characteristics both vary, so always read the specific recommendations on the label and test first on scraps of the project fabric or a comparable fabric. As a general rule, most glues marketed specifically for fabric crafting dry clear and relatively flexible and are washable—some brands recommend hand washing, while others recommend machine washing and drying. Glue customarily creates a stronger bond with cotton, other natural fibers, and napped fabrics than with synthetic or very smooth fabrics.

In general, there are two main types of washable (permanent) fabric glue on the market (Fig. 3-1):

Water-based fabric glue. Although they are water-based, these glues are formulated to hold and not wash out under normal laundering conditions. They are usually applied as a thick white fluid and are somewhat more flexible when dry than the following category of glues (but perhaps not as flexible as stitching or fusible web; see page 32).

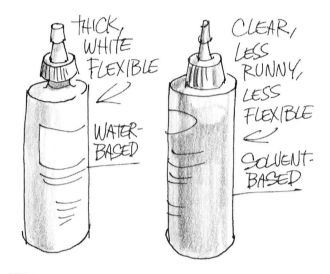

WASHABLE FABRIC GLUES:

THICK, WHITE FLEXIBLE — WATER-BASED

CLEAR, LESS RUNNY, LESS FLEXIBLE — SOLVENT-BASED

Fig. 3-1

Clean up with soap and water before the glue dries. After it dries, you may be able to remove it with denatured alcohol. Brand names of water-based fabric glues include: Aleene's *Ok to Wash-It*, Back Street *Border Patrol*, Bond *Fabric Glue* or *Wash It Again & Again*, Bond *Victory Fabric Glue* (**also dry-cleanable**), Jurgen *Jewel & Fabric Glue*, Plaid *Glu 'N Wash*, Tandy *Magique* (**also dry-cleanable**), and W. H. Collins *Unique Stitch*.

A few water-based fabric glues may need to be heat set to assure washability. Brand names include: Bond Heat & Seal, Delta *Stitchless* (also dry-cleanable), and Faultless *Fabricraft Solutions*. Apply as you would any other fabric glue, then heat set with a warm iron, following the manufacturer's instructions.

Solvent-based fabric glue. These glues are less apt than water-based glues to soak through cloth, especially on thinner fabric, synthetics, or synthetic blends, but they aren't quite as flexible when dry. They are clear in the tube or bottle, have a distinct chemical odor, and dry quickly. Clean up any excess with water while the glue is still wet. Removing the glue after drying can be difficult—acetone-based nail polish remover works on some formulas but may remove color or spot the fabric. Common brand names include: Beacon *Fabri Tac*, Creatively Yours *Fabric Plus*, Dritz *Liquid Stitch*, Magic American Chemical *Fabric Mender Magic* (**also**

CAUTION

dry-cleanable), Washington Millinery *Bridal Glue* (also dry-cleanable), and W. H. Collins *Instant Vinyl*.

Follow these general rules when fabric crafting with glue:

▲ Prewash the fabric and trim *without using a fabric softener* before gluing. Washing will remove any sizing or finish that would keep the glue from bonding well with the fabric fibers. Fabric glue often will not work on coated or "permanent-press" fabrics because there are no rough fibers and open spaces for it to cling to. Every fabric needs to be clean, dry, and at least somewhat porous for the glue to bond. Also test for any adverse reaction between the glue and a fabric made from synthetic fibers.

▲ Place a foil-covered board or shirt painting board under the fabric being glued in order to prevent bleed-through onto your work surface. On a garment or multilayered project, put the board underneath the top layer to protect the fabric below. (Fig. 3-2) To make the project base as smooth as possible, wrap and pin the rest of the garment around the board, pulling it taut (but don't stretch).

▲ The amount of glue needed is always important to know and should be considered during pretesting. Too much glue can soak through or leave a stiff appearance. Too little glue won't give the strongest possible bond. The heavier the fabric, the more glue you'll need to form the best bond.

▲ Don't delay in bonding the surfaces together after you apply a solvent-based glue. Otherwise the glue may partially dry and lose some of its strength. Conversely, for a water-based glue, you may need to let it set for a few minutes after application (until it becomes tackier), then gradually press it in place to prevent soak-through. Again, test first.

▲ When applying trims, appliqués, or emblems, apply the glue to the reverse-side edges, then position the glued item onto the fabric. If part of the glued area comes loose during laundering, just apply more glue to reattach it.

▲ First finger-press a trim or hem into position after applying glue to the edge. Then use several large books or another weighty item to hold it firmly in position until the glue sets. Weighting the fabric also forms a nice crease on a hem. (Don't forget to protect your work surface and the weights with waxed paper or another nonporous paper or layer.)

▲ Setting and drying times differ from brand to brand and depend on the temperature and humidity when the glue was applied. Many manufacturers recommend 24 hours drying time before use. Wait several days before laundering. Always follow the instructions on the label. Some manufacturers recommend turning a glue-embellished garment inside out before washing.

▲ In a hurry? Many of the fabric glues can be dried more quickly by ironing or by using a hair dryer or heat gun (used in stamp making or model crafting). Test first and always use an old press cloth or pressing sheet on both sides, just in case there's any soak-through.

Appliqué Glues

Although fusible web has long been featured for quick and easy appliqués (see page 32), fabric glue also works well and the finished appliqué center is less stiff. Back Street *Border Patrol* is marketed specifically for this purpose, although other previously mentioned fabric glues also can be used.

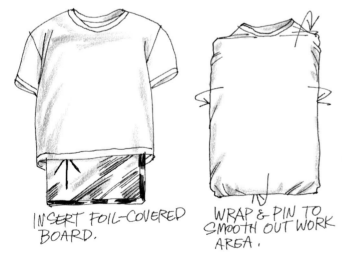

INSERT FOIL-COVERED BOARD.

WRAP & PIN TO SMOOTH OUT WORK AREA.

Fig. 3-2

Before applying an appliqué, prewash both the appliqué and the fabric you'll be putting it on. Iron the fabric to remove any wrinkles.

Place the appliqué upside down on a piece of waxed paper or foil and squeeze a thin line of glue only around the outer edges. Use the tip of the bottle or tube in a back-and-forth sketching motion, working the glue into the appliqué edges. (Fig. 3-3)

Fig. 3-3

Then finger-press the appliqué into position, beginning at the center and smoothing it out to the edges. Let the glue dry sufficiently, following the manufacturer's instructions, before wearing or laundering the garment or project.

To neatly position a large appliqué, try this trick:

1. Place the appliqué in position on the project. Holding one half of the appliqué down, fold back the other half, sandwiching a piece of waxed paper or foil between the two halves of the appliqué. (Fig. 3-4)

2. Work a thin line of fabric glue around the outer edges, following the instructions above.

3. Fold the glued half back into position on the project and repeat the same procedure for the remaining half.

Use dimensional fabric paint or a short zigzag stitch on a sewing machine to cover the raw edges of an appliqué if you want a more finished look. If

Fig. 3-4

a raw edge should lift somewhat after laundering, just reglue it.

Attaching Trims

Fabric glue is a great time-saver in attaching trims, including lace and ribbon. It also is easy to use in small detailed areas, especially when the bottle or tube has a small applicator tip. (If not, refer back to the optional fine-line applicator tip information on page 5.)

"When making bathroom window curtains from sheer white fabric, I attached a drift of silk flowers with seed and bugle beads in the centers. It took many stitches and looks nice, but was a wasted effort—I should have glued!"

Barbara Johannah, *Barbara Johannah's Crystal Piecing*

To neatly position a long section of trim in an exact spot on your project, try this idea from author Nancy Ward:

1. Place strips of masking tape so that the edges are even with the upper and lower edges of your lace or trim position, leaving the trim area untaped.

2. Working with a foil-covered board under the project fabric, bead a fine line of fabric glue along the edge of the masking tape where the upper edge of the trim will be placed.

3 Finger-press the top trim edge onto the line of glue, with the trim and masking tape edges butted. (Fig. 3-5)

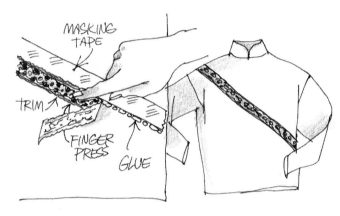

Fig. 3-5

4 When the glue is dry enough to hold the trim in a stable position, repeat steps 2 and 3 for the lower edge of the trim.

5 Cleanup is easy. Simply remove the masking tape after the glue is dry and any excess glue that may have oozed out from under the trim edges will be gone.

Quickly hem a project by turning and gluing a narrow hem to the right side. When the glue is dry, glue a row of trim or ruffled lace over the hem to hide the raw edge and the reverse side of the fabric. (Fig. 3-6)

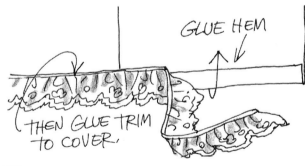

Fig. 3-6

Seaming with Glue

For doll clothes, soft sculpture, stuffed toys, Christmas decorations, and a number of other crafting projects, glue-seaming can be very effective. It's not only quick but also works well on tiny, detailed areas.

Using fabric glue, you can make just about any type of seam that can be made by stitching. A few of the glued seams used most often for crafting include:

Lapped seam. The simplest glued seam is made by lapping and joining two layers together. (Fig. 3-7) This type of seam works well on a fabric that does not fray (such as synthetic suede or felt) or for a project that will not be laundered or subjected to much stress.

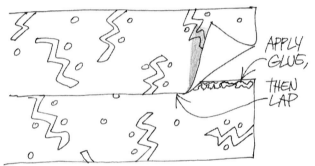

Fig. 3-7

For perfectly straight seams, place a strip of masking tape along the edge of the area where you'll apply the glue. After spreading the glue as evenly as possible, match the edge of the second fabric layer precisely to the tape edge. After joining the two sections, lift off the tape.

Make a similar seam that's less apt to fray by first glue-hemming both pieces to be joined and then lapping them to hide the raw edges. (Fig. 3-8) Use this technique only if the fabric is not too bulky.

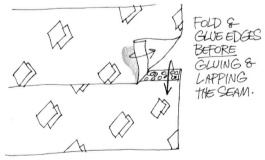

Fig. 3-8

PROJECT: Nifty No-Sew Napkins (and Placemats, Too!)

You don't have to be a talented seamstress to make attractive napkins and placemats, according to innovative author Gail Brown. Just use coordinating trim and fabric glue to finish them beautifully. (Fig. 3-9)

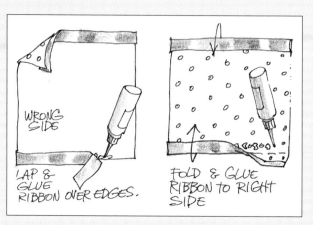

FOLD & GLUE RIBBON TO RIGHT SIDE

WRONG SIDE

LAP & GLUE RIBBON OVER EDGES.

Fig. 3-10

3. Repeat step 2 for the unfinished sides, trimming away any excess ribbon, then folding and gluing miters at the corners. (Fig. 3-11)

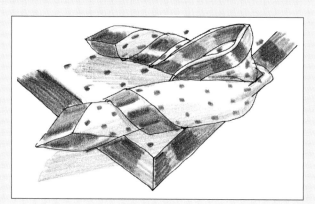

Fig. 3-9

MATERIALS NEEDED FOR SIX NAPKINS

· One yard of 54″-wide (or 1-½ yards of 45″-wide) prewashed light- to medium-weight fabric with good absorbency (such as cotton), a relatively tight weave, and no discernible right or wrong side

· 12 yards of ½″-wide prewashed grosgrain or satin ribbon (or another woven trim) to coordinate

· Washable fabric glue

HOW TOS

1. Cut six 18″ squares for the napkins and cut the ribbon into 18″ lengths (you'll need 24 pieces).
2. Working on a covered surface (see page 5), glue a ribbon section onto opposite sides of each napkin, lapping it halfway onto the fabric. (Fig. 3-10) Then glue and finger-press hems on those two sides.

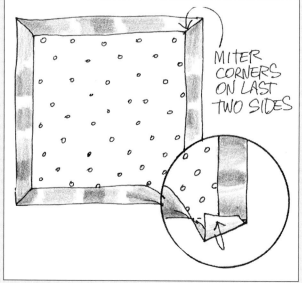

MITER CORNERS ON LAST TWO SIDES

Fig. 3-11

"For flattening glued corners, try tapping with a rubber mallet (sold at leather-crafting stores such as Tandy Leather). If you don't have a rubber mallet around (most people don't), just cover the end of a hammer with a couple of washrags."

Gail Brown, *Quick Napkin Creations*

For placemats, cut fabric rectangles 16″ by 21″ instead of 18″ square. Each placemat will require 2-⅛ yards of ribbon or trim cut into two 16″ sections and two 21″ sections. Follow steps 2 and 3 above to complete the placemats.

Simple seam. On some projects, you can glue a seam with the two pieces right sides together and hide the glued allowances on the inside. Try this technique on pillows, stuffed toys, and seasonal decorations. Just apply the glue along the right-side edge of one section, place the opposite section on top with the wrong side up, and pinch. Place the glue close to the edge so that the glued seam allowances will be narrow. (Fig. 3-12)

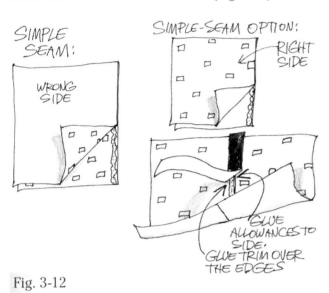

Fig. 3-12

To create a stronger simple seam, glue-seam with the two pieces wrong sides together. Glue the allowances to one side, then glue purchased trim on top to cover the edges.

Bound seam. Another way to strengthen a glued seam is to cover the allowances with bias tape. Glue a simple seam as described above, then glue double-fold bias tape over the edges. (Fig. 3-13) First glue half to one side of the allowances and let it dry. Then wrap the tape over the edges and glue the remaining half to the opposite side.

Mending with Glue

Although this is a book on crafting, here are some handy tips for quick glue repairs on your wardrobe and around your home:

▲ If a seam begins to split or pull away (especially on a delicate fabric), secure it with a strong fabric glue. On my wearable-art top with intricate pieced cuffs and yoke, the few delicate lamé fabric sections began to split at the seams where the cuffs were turned back. Because the garment is dry-cleanable, I used

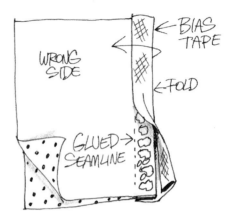

Fig. 3-13

Victory Fabric Glue and applied it with a pin to stabilize inconspicuously all of the seams that were affected. (Other dry-cleanable fabric glues could also work.) Now these seams probably will be the last places to wear out on the garment, and the glue is essentially invisible!

▲ To mend a small rip or tear, use *Fabric Mender Magic, Unique Stitch,* or *Instant Vinyl.* For reinforcement, use a section of pattern tracing paper or an unscented dryer softener sheet on the underside. Pin or tape the area together until dry.

▲ To patch a hole, apply fabric glue around the wrong-side edge of the patch and around the right-side edge of the hole. (Fig. 3-14) Place the patch (perhaps in the shape of an interesting appliqué?) over the hole and hold or weight it in place until dry.

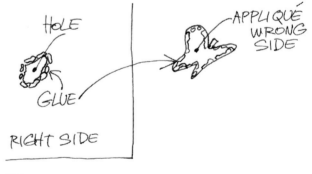

Fig. 3-14

▲ Replace stretched-out elastic in a casing by opening up a seam on the reverse side, inserting a section of new elastic, lapping and gluing the ends together, and gluing the opening closed.

FUSIBLES

Fusing is a method of applying adhesive by melting it with an iron or another heat source. It's a fast alternative to sewing, works well for applying appliqués, and has many other crafting uses, such as bonding fabric to paper, cardboard, or wood. In recent years, a wide range of fusibles has become available, including webs, films, liquids, and even a power. Fusibles can be used for almost all of the techniques previously discussed under Fabric Glues (beginning on page 26).

Fusible webs and films. These handy adhesives come in various weights (light, medium, and heavy) and forms (from yardage to packaged sheets, rolls, or tape in different widths). Some have paper backing, while others do not.

When working with a paper-backed fusible (sometimes referred to as "transfer web" because it's used to transfer an adhesive onto fabric or another item to make it fusible), preheat your iron to the setting specified in the manufacturer's instructions. Use steam only if called for. Press the unbacked side onto the item being fused, noting the time suggested in the instructions. When cool, remove the backing and fuse the opposite side to the project. See more detailed instructions under the following sections on specific applications.

Look for paper-backed fusibles under these brand names: Sew-Art International *AppliHesive* (medium weight), Therm O Web *HeatnBond Lite* (lightweight) and *HeatnBond UltraHold* (heavier no-sew weight), Aleene's *Hot Stitch* (lightweight) and *Hot Stitch Ultra Hold,* Warm Products *Steam-A-Seam* (medium weight), HTC *Trans Web* (medium weight), and Pellon *Wonder-Under* (medium weight) and *Heavy-Duty Wonder-Under.*

The *HeatnBond* products require less heat and a shorter pressing time, so they can be used with more delicate or sensitive fabrics and materials. Another innovative product in this category is *Stitch Witchery Plus With Grid* from both HTC and Dritz. It's a medium-weight fusible with a paper backing printed with a ¼″ grid so that you can create you own designs more easily, enlarge patterns from books or magazines, cut perfectly even strips for trims, and draw uniform letters and numbers. (Fig. 3-15)

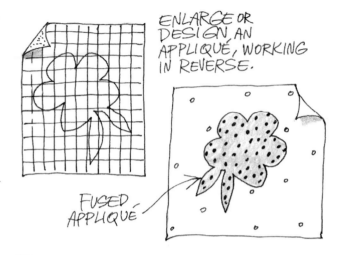

Fig. 3-15

If the fusible does not have paper backing, use a see-through *Teflon* pressing sheet, such as the Solar-Kist *Easy Way Appliqué* sheet (available in two different weights) or Clotilde's industrial-weight *Teflon Sheet:*

1. Fold the pressing sheet in half to protect the iron and ironing surface from the fusible adhesive. Place a slightly larger piece of the fusible against the wrong side of the fabric to be fused, then insert both pieces between the layers of the pressing sheet.

2. Press through all layers, following the manufacturer's instructions. Don't overpress or the bond will be weakened.

3. After cooling, lift the fused fabric (or other item) from the pressing sheet. Cut out the exact shape desired if you are fusing untrimmed fabric.

4. Place the remaining exposed side of the fusible in position on the project and fuse again, allowing it to adhere in place. (Fig. 3-16)

5. Peel any fusible residue off the pressing sheet and wipe away any excess with a soft cloth—don't scrape with your fingernail or another sharp object or you may damage the sheet.

Note: These pressing sheets are also ideal as nonstick work surfaces when using liquid glue or a glue gun.

Well-known brand names of unbacked fusible web include: Solar-Kist *Fine Fuse* (ultralight),

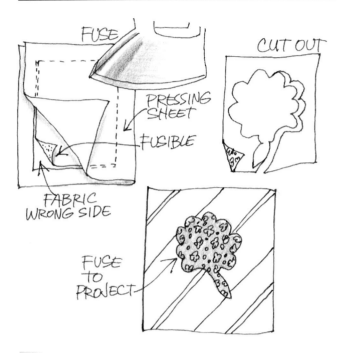

Fig. 3-16

Tuf-Fuse (lightweight), *Tuf-Fuse II* (medium weight), and *Tuf-Fuse III* (heavyweight); Pellon *Wonder-Web;* and the original *Stitch Witchery* from both HTC and Dritz.

Fusible tapes. Many of the fusible products already mentioned are also available in fusible tape form, some paper-backed and some without backing. The *HeatnBond* products, Dritz *Hem N Trim*, HTC *Trans Web*, and *Stitch Witchery* are examples. But one unique product deserves special mention: Warm Products *Steam-A-Seam* tape was designed so that the side without the paper backing will ad-here to the fabric or material *before* it is ironed in place; thus it's even easier to use. It eliminates one fusing strip and also can be repositioned *before* fusing it permanently in place.

Liquid fusibles. As an alternative to fusible web, a liquid fusible can be applied just as you would ap-ply a glue and then fused in place to form a bond. Brand names include: Aleene's *Liquid Fusible Web* and Beacon *Liqui Fuse*. Both are washable and dry-cleanable.

Powdered fusible. Aleene's *Hot Stitch Glue* is the only powdered fusible on the market at the time this book goes to press. The advantage is that a very thin layer can be applied before fusing light-weights or sheers. But in some applications, it may be a little harder to control precisely than a web or a liquid.

Other fusible products. Although they are not strictly glues or adhesives, many other fusible prod-ucts have become handy materials for crafting projects. Look for fusible fleece, trims, lace, tapes, backings and interfacings, thread, and even rhine-stones (see Chapter 4) to make your crafting easier.

When working with fusibles, keep in mind these basic guidelines:

▲ Choose the type and weight of fusible that is most compatible with your fabric and that will give the desired finished look to your pro-ject. In general, a heavier fusible has better holding power, especially for heavier-weight fabrics, but it can add stiffness or may melt through porous fabric and show on the right side.

▲ Always read the instructions on the product's package, interleafing, or backing sheet. (Fig. 3-17) Heat, steam (or no steam), fusing time, and amount of pressure required vary from brand to brand and are all important factors in achieving a strong, permanent bond. With some brands that require steam, you may want to use a damp press cloth as a tempera-ture and timing aid instead of a steam iron—when the cloth is dry, you should have a secure bond (always test first).

▲ With most fusibles, prewash all clothes and fabrics before fusing.

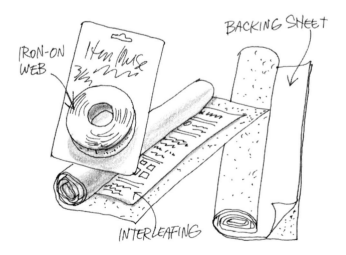

Fig. 3-17

▲ Before beginning to fuse your project, be sure to test on scraps of your fabric. Some fabrics with special finishes or irregular, heavy textures don't bond well. Also check to be sure that the fusible isn't showing through on the right side of the fabric. You may need to increase pressing time for heavier materials. Depending on the fusible brand you're using, you may need to work either on a firm surface or on a padded one—read the instructions.

▲ *Teflon* pressing sheets are essential for unbacked fusible web, and they are also a good idea for the paper-backed variety (to keep any extra bits or unaligned edges from sticking to your pressing surface or iron). Pressing sheets are nonstick and reusable, but be extra careful with pins and scissors—the lighter-weight sheets can tear easily. With some brands, a pressing sheet underneath and a nonstick iron will be all the protection you'll need. Another good option, suggested by author Nancy Ward, is kitchen (cooking) parchment paper, available in most large grocery stores. It is also reusable (any fusible residue lifts off when cool), and you can use it to turn an unbacked fusible into a paper-backed one.

▲ When fusing a large section, start in the center and work outward, overlapping the iron positions for a complete bond. (Fig. 3-18) Most manufacturers recommend putting direct pressure at each position instead of sliding the iron back and forth. Allow the fused material to cool completely before removing the paper backing or the pressing sheet.

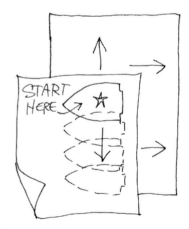

Fig. 3-18

▲ When positioning items for fusing, it's wise to fuse-baste first before fusing them permanently. Lightly touch the iron to the item being fused onto the backing or project. Or just touch the center of the pieces with the tip of the iron. This will anchor them in position so that you can be assured of the desired final effect (and adjust if necessary) before completing the fusing. An optional method of accurately positioning elements before fusing is to use Hot Tape from Distlefink Designs. It's not only heat resistant, but also it's reusable. The tape is preprinted with ¼″ markings like a tape measure, and it peels off without residue.

▲ To remove any errant fusible from the pressing surface or from an unwanted project area, use denatured (rubbing) alcohol, alcohol-based hair spray, or lacquer thinner (from a home supply or hardware store). Test first on project scraps to be sure that the fabric won't discolor or that the hair spray isn't too sticky. Wash afterward to remove any excess solvent.

▲ To clean fusibles off an iron, use a commercial iron cleaner (such as Dritz *iron-off*), an unscented sheet of fabric softener (while the iron is warm), or *WD-40* sprayed on a soft cloth or paper towel. You may also soak the cold soleplate of the iron in denatured alcohol or use hair spray on a cold iron and wipe it off when dry. *HeatnBond* recommends wiping off minor mishaps with only a clean, dry cloth. Experiment to see which method you prefer for the products you're using.

▲ All of the fusibles listed in this section are washable. Most manufacturers suggest using the delicate cycle for washing and low heat for drying. But read the instructions; some recommend hand washing and line drying. Many of the products listed are also dry-cleanable—check the instructions. If any of the fusible comes loose during laundering or cleaning, just fuse again.

▲ Store fusibles, especially those with paper backing, by rolling rather than folding. Over time, the fusible may separate from the paper, but it can still be used. Just place the fusible in position, cover with the release paper, and fuse. (Use a pressing sheet on both top and bottom, just in case your alignment isn't perfect.)

▲ A few other fusible or fusible-like films are now on the market. See the following section on repositionable glues and Chapter 7 for other bonding films.

Fusing Appliqués

Create the easiest fusible appliqués by cutting them from fusible fabric made by fusing a backing to the wrong side. Use either a paper-backed or an unbacked fusible, following the suggestions beginning on page 32:

1 Press the fusible onto the back of the appliqué fabric, following the manufacturer's instructions, and let it cool.

2 Either cut out a printed motif from the fabric design or draw a design on the backing paper or on the fusible itself. (Fig. 3-19) *Important:* A design drawn on the back of an appliqué fabric must be reversed so that it won't be backwards after you turn it over and fuse it in place.

Fig. 3-19

3 Cut out the appliqué and peel off the paper backing (if there is any). To remove a paper backing without distorting the edges of a lightweight or delicate fabric, score the center of the paper with a pin. Then gently peel the paper away from the center out.

4 Fuse the appliqué, right side up, onto the project, following the previous guidelines.

5 For added durability and a more finished look, you may want to use dimensional fabric

paint or a short zigzag sewing-machine stitch to cover the raw edges.

When your appliqué includes a number of separate but overlapping pieces, you have several options. If the fabric is lightweight, put all the pieces in position and fuse them at the same time (if several layers are involved, you may need to press for a little longer than the instructions suggest). For heavier fabrics, begin with the bottom-most pieces, fusing them in place first before lapping and fusing other pieces over them. Or, to be sure your appliqué is perfect before fusing it to the project, assemble it between two folded layers of a pressing sheet (see Fig. 3-16). After cooling, fuse the completed appliqué to the project as one piece. The Solar-Kist fusing system, with four separate weights of fusible and two grades of pressing sheets, was designed specifically for this type of application.

To save on fabric and fusible, you may choose to cut out your appliqué before fusing the back of the fabric. In that case, place the trimmed appliqué, wrong side up, on top of a paper towel, brown paper, or another kind of disposable paper. Cut out a piece of fusible slightly larger than the appliqué and place it on top. Fuse the two together, using the appropriate method for the specific fusible you're using. After cooling, just peel off the backed appliqué and position and fuse it onto the project (the extra fusible stays on the paper).

PROJECT: Western Wall Hanging

Create a delightful wall hanging for the room of a youngster who aspires to be a cowboy. Or change the motif for any other interest or occasion. (Fig. 3-20)

HOW TOS

1. Cut 1″-wide strips of the web and fuse them to the underside of the side edges of the backing rectangle, following the manufacturer's instructions. Fold 1″ side hems along the fusible's edge and press again to secure.
2. Cut one 2″-wide strip and repeat step 1 to make a 2″ bottom hem. Also press a 2″ hem at the top, but secure it with a ¼″-wide fusible strip at the raw edge only, to leave a casing. (Fig. 3-21)

Fig. 3-20

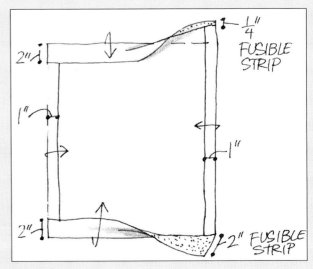

Fig. 3-21

MATERIALS NEEDED

- ½ yard of heavy muslin, felt, or suede—cut to 18″ by 24″ for the backing
- At least one repeat of a printed fabric with a cowboy or other Western motif
- ½ yard of fusible web or film
- 3 yards of medium-sized rickrack
- Fabric glue (see page 26)
- One thin 24″-long tree branch or a wooden dowel
- One printed bandanna

3. Make appliqués from the motif fabric, following the instructions beginning on page 35. Fuse them onto the backing in a pleasing arrangement.
4. From the rickrack, cut two 18″ lengths (for the sides) and two 22″ lengths (for the top and bottom). Center and glue the shorter sections of rickrack 1″ from each side edge, wrapping 1″ on each end around to the back side. For best results, lightly dot the fabric glue on each peak, then finger-press the rickrack into position. Repeat for the longer sections, positioning them 1-½″ from the remaining edges. Weight the trim down with books until the glue dries.
5. Insert the branch into the casing, tie the remaining rickrack at either side for a hanger, and knot the bandanna around one end of the branch.

Other Fusing Options

Fusibles are a speedy alternative to sewing for many crafting projects. Try these simple techniques:

Fast hemming. As on the project above, fusible strips create crisp hems quickly—just fuse to the wrong-side edge, turn the hem, and fuse again. When fusing exactly to the hem edge, be sure to use a pressing sheet or disposable paper to catch any fusible that might extend a little beyond the edge. For a less conspicuous hem, especially on some lighter or softer fabrics, position the fusible slightly away from the hem edge. (Fig. 3-22)

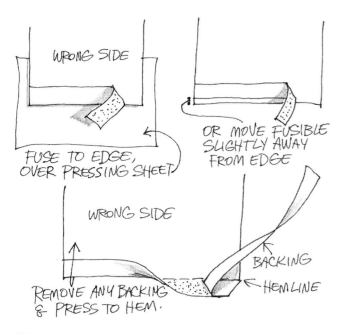

Fig. 3-22

Fused trims. Make all sorts of decorative accents fusible by backing them with fusible web or film:

1. Place purchased appliqués, crests, badges, ribbon, or trim face down on a paper towel, brown paper, or another disposable paper. (Fig. 3-23)

2. Cut a piece of fusible web or film slightly larger than the decorative item on top of it and fuse, using the method suggested for the fusible you're using.

3. After cooling, lift the decoration off the paper—the back of the item should now be completely covered with fusible.

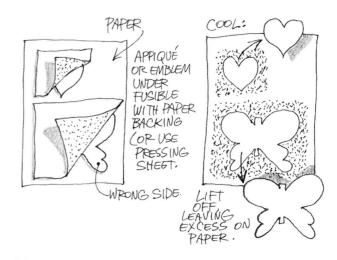

Fig. 3-23

4. If you've used a paper-backed fusible, peel away the paper now and fuse the item right side up onto your project.

Make lengths of straight-sided ribbon or trim fusible by reversing this method. Put a sheet of fusible on the pressing surface with the paper backing down (or use a pressing sheet under un-backed fusible). Pin lengths of ribbon or trim over the fusible with the edges touching and the ends extending just beyond the fusible edge. (Fig. 3-24) Cover with a press cloth or pressing sheet and fuse. When cool, the strips can be cut apart or shapes can be cut from adjoining strips, then fused into place.

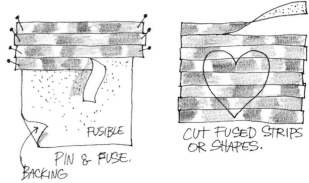

Fig. 3-24

Quick fused shapes. For making dimensional fabric shapes such as petals, butterfly or bird wings, or costume parts, creative author Yvonne Perez-Collins suggests fusing florist wire between two layers of fabric:

"Use cloth-covered wire that is green or white. Shape the wire to conform to the outline of the shape. Place it on the wrong side of the fabric that already has a fusible web pressed on. Put an additional straight or looped wire in the center area for added strength. Cover with the second layer of fabric. Press well, following the manufacturer's instructions to ensure a good bond. Trim the fabric for a clean edge. Quick crafting and no sewing!"

Yvonne Perez-Collins, *Soft Gardens*

Fused home decorating. Entire books have been published on easy (yet very attractive) no-sew home decorating. Fusibles are one of the main ingredients in this interesting revolution. Using the techniques described in this chapter, you can make curtains, window shades, pillows, lamp shades, covers for small appliances, potholders, and table linens.

Other home decorating possibilities include covering boxes, picture frames, albums, and window cornices. Repcon International (see Resources at the end of the book) features a popular selection of *Lite-Line* ready-to-decorate corrugated cardboard forms, including *Create-a-Valance* and *Tissue Box Topper,* that are perfect for covering with fused fabric. (Fig. 3-25)

CREATE-A-VALANCE

TISSUE BOX TOPPER

Fig. 3-25

REPOSITIONABLE APPLIQUÉ ADHESIVES

Several pressure-sensitive adhesive products are available to hold an emblem, appliqué, or other

"I suggest using fusible web or fabric glue to secure the rod pocket seams on the back side of a quilted cornice. It's a quick and easy finish to window treatments, eliminates stitch lines on the front of the quilted work, and holds as well as hand stitching."

Kathleen Eaton, *Stitch 'N' Quilt*

trim firmly to a project, but they've been specially formulated so that the item can be easily removed later—high technology strikes again!

Removable trims or appliqués give you several unique options. They can be removed before the garment or other item is laundered or cleaned, either because they will not withstand the same cleaning method or to better preserve them. A removable trim also gives you added flexibility—use a favorite trim on different garments or use a variety of decorative trims on one favorite garment.

Repositionable fabric glues were first marketed for crafting by Creative Potential in 1984. Its brand, *Sticky Stuff,* is now carried by Clotilde (see Resources). Other repositionable fabric glue brand names include: Glitz, Inc. *Glitz-it!,* Dritz *Insta Tack,* Jurgen *Press n Peel,* Faultless *Bead Easy Re-Apply,* Plaid *Stikit Again & Again,* and Aleene's *Tack-It Over & Over.*

Most of these repositionables are in glue form, but two, *Press 'N Wear* from Velvet Touch and *Stix-A-Lot* from Sew-Art International, are fusible bonding films that you apply as you would other fusibles. Another bonding film, *Stick & Hold for Fabric* from True Colors International, is repositionable *unless* it is heat set—then it is permanent.

Follow the manufacturer's instructions for application specifics. With these glues, you'll apply a smooth, thin coat to the back of the decorative item and let the glue dry completely before applying it to your garment or project. Apply the glue with a brush, thin sponge, or thin cardboard or plastic spreader. Store glue-backed items on nylon netting, the shiny side of freezer paper, or another waxed or slick surface (such as a contact paper or stick-on-label backing sheet) when not in use. (Fig. 3-26)

If an appliqué or other decorative item is not tacky enough or if it loses its tack after use, wipe it off with a damp towel to remove lint or just apply another coat of glue and let it dry completely be-

NETTING↑   LABEL BACKING

Fig. 3-26

fore reuse. Most repositionable glues clean up with water while still wet. After drying, wipe your fingers with vegetable oil and a paper towel. To remove dried glue from some surfaces, you'll need a solvent (such as mineral spirits or lighter fluid). Always test first in an inconspicuous spot. Be especially careful with sequinned appliqués because a solvent can remove their color.

Most repositionable glues work well on surfaces other than fabric, such as wood, plastic, glass, metal, and paper (see more paper-crafting ideas in Chapter 5). Creative Potential has tested a wide range of uses for its *Sticky Stuff*, including anchoring quilting templates and stencils, mounting needlework on a backing (it can be removed and restretched for both perfecting the alignment and for cleaning), and holding just about anything in place from dollhouse furniture and shelf paper to any items that may be earthquake-vulnerable. It also works well for making the end of a screw driver sticky as if it were magnetized, on the end of a yardstick for retrieving hard-to-reach items, and for making stickers and posting notes. Here are some of its other great uses that I've tried:

▲ Make masking tape double-faced to hold throw rugs easily in place. Apply tape to the rug edges and spread the glue on the back of the tape. When the glue dries, the rug will stay firmly in place either on a smooth floor or a carpet. It can be lifted easily for cleaning. Just peel off the tape to remove the glue. For the first time, I'm no longer frustrated by "traveling" throw rugs!

▲ Don't bother to attach shoulder pads or even a *Velcro* strip to each garment. Simply put masking tape across shoulder pads and coat it with pressure-sensitive glue. When the glue dries, the pads will stay firmly in position and can be removed for cleaning or to use in other garments. Store with a piece of netting over the glue and reapply more glue if the tape collects fibers and loses its stickiness.

▲ For a soft and flexible leather hem, apply the glue to the entire hem area. (Fig. 3-27) When dry, fold the glued area in half, glued sides together, for a permanent bond.

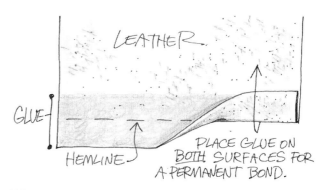

LEATHER

GLUE

HEMLINE   PLACE GLUE ON BOTH SURFACES FOR A PERMANENT BOND.

Fig. 3-27

THREAD AND FIBER SEALANTS

Often referred to as "seam sealants," these glues can be used to seal any loose threads or fabric edges. They're commonly used for serging projects to permanently seal the loose ends of stitching so that the garment won't ravel. Sealants also serve the same purpose for any thread or yarn ends. Another tip: They're great for stopping runs in hosiery and for repairing the ends of frayed shoelaces.

Some brands dry softer than others, so test first before using sealant in a place that will be in direct contact with bare skin—it may be scratchy. Use denatured alcohol to remove unwanted drops from your project. Look for these common brand names: Dritz *Fray Check*, Bond *No Fray*, Brohman *No-Fray* (available from Clotilde, see page 97), and Aleene's *Stop Fraying*.

PROJECT: Quick Change Seasonal Sweatshirt

Make a super-easy and eye-catching top to casually celebrate any season or special event. (Fig. 3-28)

MATERIALS NEEDED

· An undecorated purchased sweatshirt in any color
· Purchased appliqués—one or more for each season or event, in colors that complement the sweatshirt
· Repositionable appliqué glue
· Nylon netting
· Waxed paper

HOW TOS

1. Apply glue to the back of each appliqué piece, following the manufacturer's instructions, and let it dry completely. Work over a sheet of waxed paper for easiest cleanup.
2. Position appliqués on the sweatshirt that are appropriate for the next season or event. Store the appliqués on nylon netting or freezer paper when not in use.
3. Don't leave any glued appliqué on the sweatshirt for more than a few days, and always remove the appliqués before laundering or cleaning. Reapply more glue if the appliqués lose their tackiness.

Fig. 3-28

PROJECT: Terrific Tassels

Create simple decorative tassels to embellish accessories or any home decor item. (Fig. 3-29)

Fig. 3-29

MATERIALS NEEDED

· Yards of decorative thread or cording—from 6 yards for a heavier variety up to 14 yards for a lighter weight (you'll also need more if you want a fatter tassel)

· Liquid sealant

"If you have a serger, quickly serge yards of thread chain to make attractive textured tassels. For a delicate look, use rayon, lingerie, or fine metallic thread. Try pearl cotton, pearl rayon, or crochet thread for a heavier tassel. Rayon gives the softest, swingiest appearance."

Naomi Baker, *Know Your Serger*

HOW TOS

1. Loosely wind the thread or cord over a firm piece of cardboard cut to the length you want the finished tassel—don't stretch. The more you wind, the fuller the tassel will be. (Fig. 3-30)
2. Cut a 6″ strand for the tassel hanger and tie it around all of the thread or cord at one end. Loop one end and tie again in the same spot, leaving tails about 1″ long. Dab sealant on the knots and all of the strands at the opposite end and allow it to dry.
3. Cut the sealed ends evenly and lift the tassel away from the cardboard. Wrap another strand around the tassel, about ¾″ from the tied end, catching the loop tails. Secure the loose ends of the wrap by tying a knot, trimming the ends to about ¼″, adding a drop of sealant, and tucking the knot under the previous wrapping.

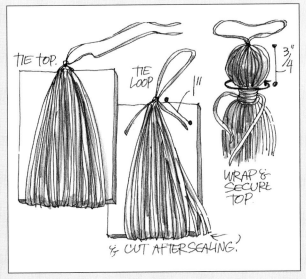

Fig. 3-30

4

Embellishing Just About Anything

- ▲ Glitz Glues
- ▲ Paper and Fabric Laminates
- ▲ Transfer Mediums

In previous chapters, you've already read about some popular embellishment glues and techniques, including découpage (page 15), appliqués (pages 27 and 35), and repositionable adhesives (page 38). But there are many other glues and adhesives that have been developed and marketed specifically for embellishments—using gems, foil, glitter, and more—on many types of crafting projects. Still others are formulated to laminate paper and fabric to a variety of surfaces.

When deciding which glue to use for many of the techniques described in this chapter, remember that a number of the glues listed previously may also be excellent. One major example is the list of fabric glues in Chapter 3—read the labels. Most of these glues will work just as well bonding ornamentation to fabric as they do bonding fabric to fabric. And in addition to being washable, several are dry-cleanable, too. **Note:** Because dry-cleaning solvents vary, always test a sample before dry-cleaning your finished project.

GLITZ GLUES

One of the most popular and exciting crafting alternatives in recent years is fast and easy "wearable art." Using ready-made "blanks" (unadorned causal wear such as sweatshirts and T-shirts) or a garment you've constructed yourself, you now have so many embellishment options that you may find it difficult to decide which one to choose.

As with the other glue categories previously discussed, performance varies from brand to brand. Read the instructions carefully. If you'll be washing the project, be sure to prewash it (without fabric softener) before applying any ornamentation. If you'll be dry-cleaning it, read the label carefully (and test if in doubt) to be sure that the glue will hold. When you're gluing any decorative object to your project—from the largest button or jewel to the finest glitter—you'll want to be sure it stays in place permanently.

Anyone who becomes an enthusiast of embellished garments and other fabric items learns that dimensional fabric paint is another popular craft medium. If you work with these paints, available in a wide range of colors and varieties, you'll find that they also can be used to bond decorative objects to your project. Jones Tones *Plexi 400 Stretch Adhesive* is essentially a dimensional fabric paint that has been formulated to dry clear.

Most of the glues discussed in this chapter can be used to glue embellishments to wood, glass, metal, and plastic as well as to fabric. For an even stronger bond between two nonporous materials, such as between jewels and metal jewelry findings, see the suggestions in Chapter 6. When you're working with paper or other surfaces where a strong washable bond isn't necessary, refer to Chapters 2 and 5 for other glue options.

Jewels, Buttons, and Other Trims

One of the easiest ways to turn a plain garment into a show-stopping one is to embellish it with creativity and flair. (Fig. 4-1) The savings can be outstanding also. Premium prices are charged for everything from glitzy T-shirts and tennis shoes to the finest jewel-encrusted gowns. Gemstones, pearls, and myriad other trims are used to fashionably ornament all sorts of clothing projects as well as many other craft items.

Fig. 4-1

Several types of glues and adhesives are used for these kinds of craft projects:

Solvent-based adhesives. In addition to the solvent-based fabric glues mentioned on page 26 and such high-tech glues as Bond *Victory 1991,* one other glue is marketed specifically for attaching jewels to fabric—Creatively Yours *Jewelry & More.* Each of these glues are fast-drying and also strong enough for jewelry making (other jewelry adhesives are discussed in Chapter 6).

Water-based glues. Most of the glues used to embellish garments are water-based and therefore are washable and nontoxic. These glues are often white when applied, but they dry clear. For other good options, see the fabric glues listed on page 26.

The wide variety of brand names in this category includes: Aleene's *Flexible Stretchable Fabric Glue* and *Jewel-It,* Beacon *Gem Tac,* Bond *Jewel Glue,* Delta *Jewel Glue* **(also dry-cleanable),** Emson *ML2000,* Jenny Jem's *Confetti Glue,* Jones Tones *Plexi 400 Stretch Adhesive* **(also dry-cleanable),** Mark Enterprises *Mrs. Glue For Glitter & More,* and Plaid Jewel Glue.

If you plan to dry-clean your project, use Delta *Jewel Glue* or one of the dry-cleanable fabric glues. Also be sure to use dry-cleanable jewels and trims.

Follow these general instructions for applying most jewels and stones to fabric:

1 Prewash the garment without using fabric softener. If the fabric is coated or has a special finish, test first on an underside seam allowance to be sure the embellishment will hold after gluing.

2 Mark the positions where you want to place the embellishments, using chalk or a disappearing marker. (Or use iron-on transfers as guides to bead placement.)

3 Cover your work surface with waxed paper and use a shirt board (see Fig. 3-2) when working on a garment. Apply a small puddle of glue (slightly larger than the bead or jewel) in the place you want to position it. For large jobs, use a glue syringe (see Fig. 1-8).

4 Place the bead or jewel directly over the glue. When gluing many jewels or beads, the pros at Bond recommend using a ball of *E-Z-Tak* reusable adhesive (see page 81) on the end of a manicure orange stick to pick up each item and position it accurately. Then gently push the embellishment into the glue, using your finger or a toothpick. The glue should come up and around the sides of each stone to hold it firmly in place. (Fig. 4-2)

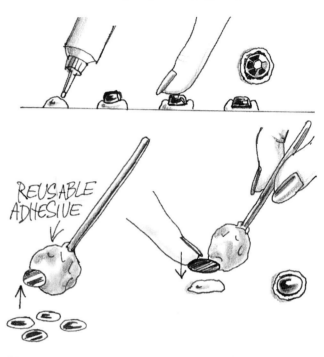

REUSABLE ADHESIVE

Fig. 4-2

Note: One exception to this general application technique is Beacon *Gem Tac*—use only a small amount, just enough to coat the back of the stone.

5 Follow the manufacturer's instructions for recommended drying times before wearing and washing the garment.

If you're gluing on larger items such as buttons, you probably won't need to surround and embed the item in glue—just use enough to hold it securely after it dries. When gluing a length of trim or cord, follow the above procedure, but draw a line of glue where you want to position it and pin or tape the trim in place until it dries. See other trim applications techniques beginning on page 28.

Problems? If the glue soaks through your fabric, you're either using too much or applying too much pressure. If the ornaments don't hold after laundering, you may not have used enough glue, the garment may not have been prewashed properly, or the ornament may have been dirty.

Fusibles. Both fusible webs (page 32) and liquid fusibles (page 33) can be used to apply ornaments to fabric. (On other surfaces, glues are usually the advisable choice.) See page 35 for instructions on fusing appliqués.

When fusing a glitzy sequinned or jeweled appliqué to your project, consider these additional pointers:

▲ The heat needed for fusing an embellished ornament can melt some acrylic items, so use care! One alternative is to use a fusible that requires low heat, such as one of the *HeatnBond* products listed in Chapter 3.

▲ Place a doubled towel or a layer of batting on your pressing area if you are working with an embellished appliqué that has an uneven surface. Because you'll be fusing with the appliqué right side down, the towel or batting will cushion it and help form a better bond.

▲ When applying fusible to the back of a jeweled appliqué, follow the instructions for fused trims (page 37) or sandwich the appliqué upside down with fusible web on top between two layers of a *Teflon* pressing sheet. You may need to use several layers of fusible to achieve the best bond.

▲ Press intermittently and carefully, following the manufacturer's instructions, checking that the heat is not damaging the appliqué.

▲ Fuse the appliqué in position on the project, following the same instructions and precautions. Depending on the appliqué and the fusible you're using, you may need to fuse from both the top and undersides for a strong bond.

Rhinestones and Mirrors

The jewel glues mentioned specifically in this chapter will work well for gluing rhinestones to your projects. Other glues, including many of those listed in Chapter 3, are also effective. But be careful when using any solvent-based glue for rhinestones—test thoroughly first to be sure that the reflective backing doesn't crack or become cloudy.

Rhinestones with a special iron-on backing are also available and, until recently, had to be positioned on the right side of your fabric and blindly ironed on or set with a special tool from the underside. Now there's a good option for applying adhesive-backed rhinestones from the right side so that you can see exactly where you're putting them—it's a hot tool with a special tip.

Jehlor Fantasy Fabrics, because so many of their customers make skating and other elaborate costumes, has developed a hot knife (high-heat cutting tool) with two special adapter tips that pinpoint heat down through the stone itself to set the adhesive on the back. The company also offers a variety of *Hot-Fix* stones in addition to rhinestones. If you're doing a lot of this kind of work, this new method can be a real time-saver. Clotilde also offers a *Gemtool* kit that's used to adhere adhesive-backed jewels to fabric.

Bead Art

Intricate hand beading using seed beads has long been a beautiful form of decorative art. But unless you have good eyes and a good deal of time, seed beading may be more work than you have patience for. With today's excellent glues, you can quickly get the same effect. (Fig. 4-3)

Fig. 4-3

PROJECT: Button-Ornamented Vest

An antique pearl pin became the inspiration for this charming vest designed by Karen Dillon. Update a vest that you're tired of, cover stains or tears in an old favorite, or perk up a plain new one. (Fig. 4-4)

MATERIALS NEEDED

· One unembellished vest, ready to wear
· Buttons, beads, and other decorative accents
· A glitz or fabric glue

HOW TOS

1. Scatter all of the decorative items onto the front of the vest and rearrange them until you're pleased with the appearance.
2. If you have included a pin in your design, just pin it in place. Then place a foil-covered board or shirt board underneath, lift each button one at a time, add glue to the back, and return it to the same position. Press or tap lightly to be sure it adheres.
3. Following the glue manufacturer's instructions, wait until the glue dries adequately before wearing the vest, and wait until the glue sets permanently before dry-cleaning or laundering.

Fig. 4-4

Using a washable glue that adheres to a soft, flexible fabric, you can permanently add seed beads to a garment or accessory in minutes. (You can add seed beads to other surfaces, too.) Most of the water-based glitz glues listed in this chapter can be used for bead art, but they must be stretchable if you are using a fabric that will stretch. One good option is Delta *Jewel Glue,* which is also dry-cleanable.

One relatively recent entry into the bead art field is Faultless with its *Bead Easy* products. Faultless produces an entire color line of miniature beads (no holes), as well as two adhesives (one that dries with a glitter or iridescent sheen) and a *Clear Top Coat Adhesive* sealer.

To apply bead art:

1. Work over a shallow pan or tray to catch any excess beads.

2. Prewash (without fabric softener) any project that will be laundered later. If necessary, press when dry to remove wrinkles. Then, if you're working on a garment, place a foil-covered board or shirt board under the top layer.

3. Apply glue to the areas you want to bead, following the manufacturer's instructions. For *Bead Easy,* you'll need only a thin coat. For other glues, use a more generous amount—test first.

4 Sprinkle a generous amount of beads over the glued area until it is completely covered. Tamp the beads into a single layer by first covering the area with a sheet of waxed paper and then using a rolling pin, the adhesive bottle, or your fingers. Immediately remove the waxed paper.

5 After the glue dries, add a protective top coating to the top and sides of all the beads.

6 With the *Bead Easy* products, heat setting is necessary if you will launder the project. Press with a warm iron on the reverse side for 10 to 15 seconds (don't forget to use a press cloth).

7 On a washable project, follow the manufacturer's instructions regarding drying time and laundry methods. It's better to err on the side of caution, letting the work dry as long as possible and treating it as very delicate.

When doing bead art, also consider these tips and suggestions:

▲ Use a brush to apply the glue to larger areas or if the bottle does not have a fine-point applicator tip. Wash the brush in warm water before the glue dries.

▲ To precisely control the placement of the beads, use masking tape to outline the design area before applying the glue. Remove the tape after applying the beads and letting the glue dry. Or, use a stencil when applying the glue, lift it off, then apply the beads.

▲ Transfers are available for easy application of more elaborate designs. Or, highlight sections of a printed fabric.

▲ When working on a large project, complete one area at a time and let it dry before moving to another area.

▲ Create different effects by scattering beads on a project or building up drifts of beads for a special effect. (Fig. 4-5)

▲ When beading a painted surface or one that contains another craft application, let the surface dry completely before applying the beads.

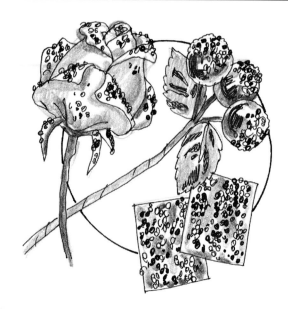

Fig. 4-5

▲ If your project is temporary or won't be handled during use, you may choose to skip the sealer coat. The beads won't be affixed as permanently, but in these cases the extra strength won't matter.

▲ Pick up and save extra beads using a rolled cone of waxed paper or a funnel. If some glue residue remains on them, wash it off before it dries, using a fine-gauge kitchen strainer and warm water.

▲ Even with a dry-cleanable glue, spot-clean bead art projects as much as possible instead of completely cleaning them.

▲ If some beads do fall off your project, reapply more using the same procedures.

Glitter

Glittering is another popular accent used for a wide range of projects. It's available in a variety of sizes from fine to course, metallic or opalescent, and in a wide range of colors. (Fig. 4-6)

Apply glitter over wet glue or dimensional fabric paint, following the manufacturer's instructions. Although most of the water-based glitz glues work well for glittering, when working on fabric that has lots of stretch, use a glue that will remain flexible after it's dry, such as Jones Tones *Plexi 400 Stretch Adhesive* or Aleene's *Flexible Stretchable Fabric Glue*.

▲ After the glue dries completely, use a soft-bristle toothbrush or other soft brush to remove the excess glitter. To conserve, brush the glitter onto a sheet of paper and roll the paper into a funnel to return it to the container. Or for quicker cleanup, pick up the excess using a lint roller with replaceable masking tape.

▲ Before laundering, wait for the glue to dry completely, then follow the manufacturer's instructions. Gentle hand washing and line drying are usually recommended.

Sequins

Sequins come in several types, but basically they are the same material as glitter. When using the smaller sprinkle sequins (confetti), apply them as you would glitter. (Fig. 4-7)

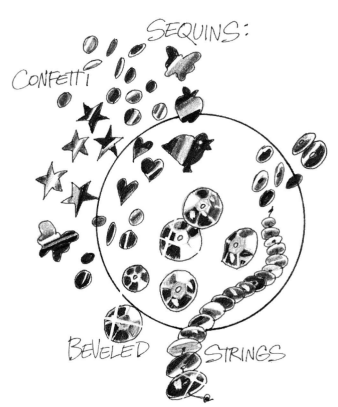

Fig. 4-7

For larger sew-on sequins with beveled or curved surfaces, apply them as you would a jewel or button (see page 44). Make sure that the glue comes up through the center hole of each one to anchor it securely. For strings of sequins, follow the instructions for applying trim (page 28).

Fig. 4-6

Mark Enterprises *Mrs. Glue For Glitter & More* is a washable, flexible, water-based glue designed specifically for applying glitter. Another popular glitter glue is Jenny Jem's *Confetti Glue*, which was developed for gluing *Mylar*. It's a good choice for gluing glitter to a *Mylar* shape or for gluing two pieces of *Mylar* together.

When applying glitter, follow the general instructions for bead art beginning on page 46. Also consider these other suggestions for successful glittering:

▲ When using more than one color of glitter on a project, work with only one color at a time to avoid any mixing.

▲ Gently blow away any excess glitter from areas still to be worked on.

Foiling

Foil can be used to embellish your crafting projects, from garments and accessories to glass, wood, and metal. Use it also to enhance hot glue gun designs (see Chapter 8).

For fabric foiling, you have the choice of using a heat-set product, such as *HeatnBond* (page 32) or *Stitchless* (page 26), or a washable fabric or glitz glue that doesn't require the heat-setting process.

Although many of the glues already mentioned in this chapter can be used for foiling (Jones Tones *Plexi 400 Stretch Adhesive* is a good example), other products are marketed specifically for this technique. Plaid *Liquid Beads Dimensional Bond* is one brand, available in *Press & Peel Foil* kits with the foil and sealer included. The key to a good foiling glue is that the glue must remain sticky when dry so that the foil can easily adhere to it.

Follow these steps to apply foil to fabric:

1 Prewash (without fabric softener) any project that will be laundered later. If necessary, press when dry to remove wrinkles. Then, if you're working on a garment, place a covered board or shirt board under the top layer.

2 Following the manufacturer's instructions, use the glue to write, draw a design, or embellish a design already on the fabric. Put the project in a dry, dust-free place until the glue is dry and completely clear.

3 Place the foil with the color side up (dull side down) over the glue and press gently but firmly. (Fig. 4-8) Then peel away the foil sheet. Press and lift, using different colors if desired, until all the glue is covered.

4 For extra durability, apply a sealer over the foil, following the manufacturer's instructions.

5 Launder according to the product instructions. Don't dry foiled projects in a dryer or iron over the foil.

Although most of the foiling glues already mentioned also work well on hard surfaces, several other products have been developed specifically for foiling projects other than fabric. Aleene's *3-D Foiling Glue*, Anita's *Foil Adhesive* (with a pinpoint applicator) and Duncan *Easy Foil Medium* fall into this category and are usually used for decorative home accessories.

Look for manufacturer's project instructions

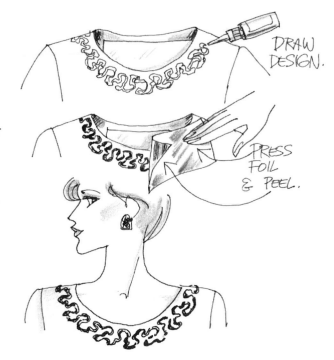

DRAW DESIGN.

PRESS FOIL & PEEL.

Fig. 4-8

and kits that provide much more information on foiling techniques, including metallic leafing and combing.

PAPER AND FABRIC LAMINATES

Now it's possible to embellish practically any surface with an interesting multicolor design or print. Both paper napkins and fabric can be used in a number of creative ways.

Napkin Appliqué

In Chapter 2, you read about the art of découpage. Some creative manufacturers have taken découpage one step further, in which thin paper napkins (or tablecloths) are affixed to surfaces varying from fabric to plastic. (Fig. 4-11)

After you "appliqué" a printed napkin cutout to a surface, the lighter background colors practically disappear and the colored design seems to be painted right onto the project. Well-known brands used for this technique are: Aleene's *Paper Napkin Applique Glue*, Delta *Naplique Paper Applique*, and Faultless *Fabricraft Solutions* (the latter two require heat setting).

PROJECT: Fancy Foiled Box

Add a stunning embellishment to any wood box, using the Metallic Combing Technique, recently introduced by Duncan. (Fig. 4-9)

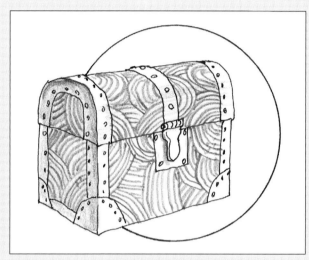

Fig. 4-9

MATERIALS NEEDED

- Painted wooden box with lid
- *Duncan Easy Foil Medium* (adhesive)
- Metallic foil
- A comb (or other blunt object for texturing)
- Sealer (matte or gloss, depending on the effect desired)
- Masking tape

HOW TOS

1. Using masking tape, outline the areas to be foiled on the top and sides of the box. (If desired, paint these areas with a different color and let it dry first.)

2. Using the brush, apply the adhesive to the outlined areas. The thicker the application, the more texture you'll be able to create with the comb or other tool.

3. Create a textured effect in the wet adhesive before letting it dry. (Fig. 4-10) If you're not happy with the design, brush it away and try again. Even brush strokes can produce an interesting effect. Remove the tape and let the adhesive dry until clear. Wash out the brush with soap and water before the glue dries.

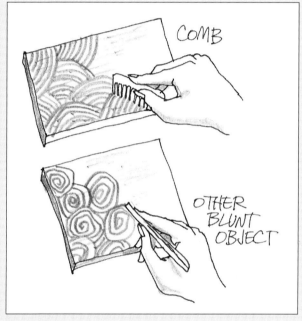

Fig. 4-10

4. Place the foil, shiny side up, over the dried adhesive. Apply pressure with your fingers—the more pressure, the more coverage you'll get. Then peel away the foil sheet. For better coverage, apply additional foil from the sheet. If you're still not satisfied, begin the process again, adding more adhesive and then more foil.

5. Apply sealer to the entire foiled surfaces and allow it to dry.

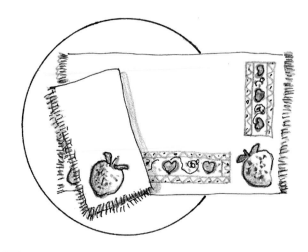

TRANSFER MEDIUMS

Some liquid fusibles or heat-set fabric glues (see page 26) can be used as a transfer medium to transfer a photo or printed image to fabric. (Fig. 4-12) Delta *Second Impressions* kit, containing *Stitchless* Washable Fabric Glue, a roller, sponge, and complete instructions, is one product that's marketed for this technique.

The procedure involves photocopying the image you want to transfer, either in black and white or in color. Because the image will then be reversed on the fabric, you can also copy onto acetate, turn the image over, and recopy—this is especially important for letters and words.

To apply a napkin appliqué, follow the manufacturer's instructions. In general, you'll apply glue to the project and add a single ply of a printed napkin cutout. Then you'll apply a second layer of glue on top of the napkin. To finish the design, you may add embellishments of jewels, fabric paint, or glitter.

Treat fabric projects as delicate washables (after the recommended drying time) and line dry. When using napkin appliqué on other decorative projects, apply a sealant when dry to protect against dirt and dust. If the project will be exposed to the elements, use a waterproof sealer.

Laminating

Soft fabric or pliable paper (napkins, tissue, or gift wrap) can be applied to either side of a project to transform it from ordinary to spectacular. When laminated to the top, the fabric or paper is like an all-over napkin appliqué. When laminated to the reverse side of clear glass or plastic, it becomes a "reverse collage" and looks like an interesting mosaic.

Because most craft laminating is done on uneven surfaces, the fabric or paper is usually cut into sections. Even silk flowers or other lightweight accents can be added effectively, especially on a reverse collage.

Products designed and marketed for laminating include: Aleene's *Reverse Collage* Glue and Beacon *Liquid Laminate*. Anita's *Découpage Glue* (for wood, metal, glass, and plastic) and J.W. Etc.'s *First Step* (for laminating and sealing fabric to wood) are other options.

Following the manufacturer's instructions, apply the glue to the photocopied image, place it glue-side-down on the fabric, and roll it firmly in place. When dry, heat set it with a dry iron, let it cool, and then wet and gently rub away the paper, leaving the image on the fabric. Then sponge or brush on diluted glue when dry to coat the transfer, sharpen the image, and protect it during wearing and laundering.

Using essentially the same technique (Delta calls the method *Sheer Impressions*), you can transfer your image to the reverse side of sheer fabric. Here the transfer will not be reversed, and it will be protected by the layer of fabric during

PROJECT: Decorated Flower Pot

Make a one-of-a-kind flower pot using your imagination and a laminating adhesive. (Fig. 4-13)

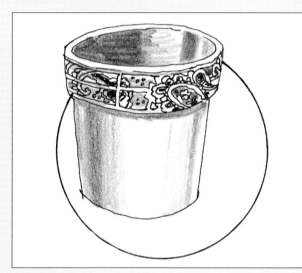

Fig. 4-13

MATERIALS NEEDED

- One clay pot or an enameled pot with a clay rim
- Print napkins with an all-over design
- Napkin appliqué glue or Beacon Liquid Laminate
- Paintbrush
- Sealer or spray gloss

HOW TOS

1. Cut motifs or sections of the napkins to cover the pot or the rim, overlapping as necessary. Then peel off the top ply of the cutouts.
2. Apply the adhesive to the surface to be laminated, using the brush.
3. Position the cutouts over the glue, applying more glue under the lapped edges. Press out any air bubbles. Wash out the brush with water while still wet.
4. After the project dries, finish it with sealer or gloss, letting the surface dry between coats.

use. Of course, the image will be slightly less sharp because you're seeing it through the sheer layer.

Aleene's *Transfer-It* is another transfer-medium brand that produces a very clear image. It can be used without being heat set, if desired.

Transferred images can be gently washed but should not be dried in a dryer or dry-cleaned.

5

Crafting with Paper

- ▲ Mucilage and Pastes
- ▲ Rubber Cements
- ▲ Glue Sticks
- ▲ Clear Paper Glues
- ▲ Repositionable Paper Glues
- ▲ Spray Adhesives

Paper crafts have long been recognized as an excellent way to introduce children to art. Paper projects are fast, inexpensive, and offer a fun chance for self-expression. But paper crafts are not just for kids—stationery, collages, and papier mâché are just a few of the artistic results you can accomplish.

The white, tacky, and clear glues listed in Chapter 2 all work well for gluing paper. But there are other glues you can use for paper crafts as well. In many cases, several different types of glue work equally well to complete the same project. When you make your selection, consider safety (especially important when working with youngsters), cost, ease of use for the particular project (drying time, spreadability, and cleanup, for example), and the end results.

MUCILAGE AND PASTES

Few of us who went to grade school in the United States haven't used paste and mucilage. These glues are usually natural-based and all are non-toxic—plus for many of us, just the smell brings back many memories. (Fig. 5-1)

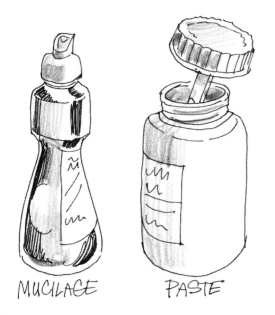

MUCILAGE   PASTE

Fig. 5-1

Mucilage and paste are used mainly for kids' crafts and in the classroom. Although not as strong as many other glues, they work well for simple projects. Some of the less expensive white and school glues (see Chapter 2) are now often preferred for general classroom use.

Mucilage

Traditionally available in a shaped bottle with a rubber applicator tip, mucilage is marketed by companies such as Borden (Elmer's), LePage's, and Ross. It dries relatively quickly and generally causes less rippling on lightweight paper than a white or school glue.

Some brands of mucilage can stain surfaces and dry hard, so test first before using one on a project that might be affected.

School Paste

White paste (sometimes called "library paste") causes about the same amount of rippling on lightweight paper as a white or school glue and has a similar medium-length drying time. It can also lump and crack, so don't count on it for longevity.

Spread paste on the project with an applicator such as a wooden craft stick (some brands come with a built-in spreader attached to the lid). Common paste brands include: Crayola, Elmer's, LePage's, and Ross.

Wheat and Art Paste

Wheat and art pastes come in a dry powdered form and must be mixed, but they're excellent for many arts and crafts projects, from finger painting to papier mâché and collage. Wheat paste is also called "wallpaper paste," and, as the name implies, it's often used for hanging wallpaper.

Completely nontoxic, wheat and art pastes are marketed under brand names such as: Evans *Banner Wheat Paste, Golden Harvest,* and Ross *Art Paste* (formerly Pritt Art Paste).

Papier mâché is one of the most popular paper crafts today and may be made by anyone from children to fine artists. It was first developed in China during the second century. Now there are so many types, techniques, and decorative options for papier mâché that entire books are available on the subject. Jewelry and decorative home accessories are the most frequently made projects. (Fig. 5-2)

Papier mâché is made from paper, water, and some kind of adhesive. Although white, tacky, and other glues (even flour and water boiled into a paste) can be used, wheat paste is a good choice because it's inexpensive and often has a fungicide to prevent any mold from forming if the drying process is slow. Any type of paper can be

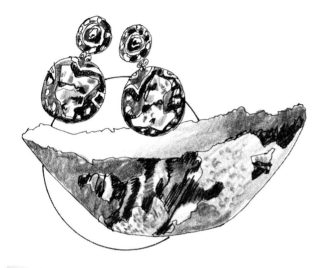

Fig. 5-2

used—the type will change the appearance of the finished project.

There are two general methods for making papier mâché:

▲ **Pulp method**—Small bits of paper are soaked, boiled, and strained to form a soft, wet pulp. Glue or paste is added, then the mixture is shaped in a mold or over a form (even cardboard) to dry. (Fig. 5-3)

Fig. 5-3

▲ **Layering method**—Squares or strips of paper are randomly layered onto a surface with the edges overlapping. The paper is either soaked in the glue mixture or the glue is brushed on the surface. Each piece of paper is applied

and then more glue is brushed on top of that. Almost any surface can become a mold for layering—even a balloon to make a rounded hollow shape.

After the papier mâché is dry, it will be light-weight, hard, and strong and can be painted and/or coated with a sealer (either matte or gloss). Ornaments or foil (see Chapter 4) can be added—the options are endless.

RUBBER CEMENTS

Rubber cements include solutions of either natural or synthetic rubber. Applied with a brush (often built into the container lid), rubber cement is strong and quickly bonds paper and cardboard. It also works well for leather, suede, and synthetic suede. Although not necessarily the best choice for an all-purpose adhesive, rubber cement will also bond paper to metal, glass, and wood. (Fig. 5-4)

Fig. 5-4

"Rubber cement is often used to attach an un-mounted stamp to a block. It's impervious to water but not to solvent-based products (inks and cleaners), so wipe these products off after use."

Nancy Ward, *Stamping Made Easy*

Often the choice of commercial artists for mounting and graphic arts, rubber cement dries quickly, gives a flexible bond, and won't cause wrinkling. But be careful because rubber cement

PROJECT: Papier Mâché Bracelet

Create fashion jewelry to match any outfit or to give as an artistic gift. Using the same techniques, you can also make earrings and pins by gluing jewelry findings to the backs of flat pieces. (Fig. 5-5

Fig. 5-5

MATERIALS NEEDED

· A section of a cardboard mailing tube, cut to the width desired (or an old, lightweight bracelet you no longer wear)—it should fit loosely over your hand because you'll be adding layers to all surfaces

· Scraps of gift wrap, tissue, or any other flexible paper in a print or color you want the finished bracelet

· Wheat paste

· Sealer

· Stones or jewels

· High-tech (see page 22) or solvent-based (see page 45) adhesive

HOW TOS

1. Mix the wheat paste according to the manufacturer's instructions and put it in a small flat dish.
2. Cut the paper into small strips. Dip each strip into the glue mixture and pull it between your fingers to squeeze out any excess.
3. After applying glue, smooth each strip onto the bracelet base, overlapping the edges and covering the entire surface. Apply at least six layers.
4. Let the bracelet dry, hanging or turning it occasionally so that it dries evenly.
5. Spray or brush on several coats of sealer, letting it dry between coats. Using the adhesive, glue on the stones or jewels as desired.

often discolors photos and, with age, may discolor some paper or the bond may weaken.

Cleanup is easy: after use, just rub away any excess with your fingers. For large projects or for corners and small areas, Best-Test markets an eraser called Pik-Up to make the job easier. Special solvent and thinner are available for big jobs or for extensive use.

Well-known rubber cement manufacturers include: Best-Test, Elmer's, LePage's, Ross, and Sanford. Their products are marketed either as paper cement or rubber cement.

Apply rubber cement one of these three ways:

▲ Cover both surfaces and let them dry before aligning them. Don't press, because a little ad-

justment can be made before burnishing (rubbing the pieces firmly together with a burnishing tool or another hard, rounded-edge object, such as a wooden spatula). When alignment is crucial, use a slip sheet (a sheet of clean paper) between the coated surfaces, removing it a small distance at a time as you achieve the exact alignment. The dried rubber cement will not stick to clean paper. (Fig. 5-6)

▲ Apply rubber cement only to one surface and position the piece immediately, before it dries. In most cases, this creates a temporary, repositionable bond. For more strength, you can let the glue dry and then burnish the piece down.

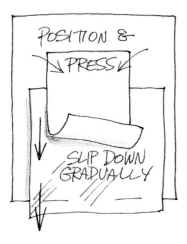

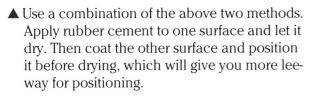

Fig. 5-6

Fig. 5-7

▲ Use a combination of the above two methods. Apply rubber cement to one surface and let it dry. Then coat the other surface and position it before drying, which will give you more leeway for positioning.

Rubber cement also can be used as a resist in painting. Simply coat an area you want to protect from the paint, work around it, then roll off the glue when the paint is dry.

Other products in the rubber adhesive category include "latex adhesives," which are water-based and nonflammable. One good example is Lakeside Plastics *Tri-Tix,* a rubber cream glue that can be used for many all-purpose jobs (like a white glue) but is flexible when dry and rubs off for cleanup. Two other nonflammable options are *Rubber Cement GluTube* and *Quilters' GluTube,* which have a built-in ball tip. They're good for small jobs and for simplified quilting techniques.

GLUE STICKS

Safe and nontoxic, glue sticks are economical and easy to use for gift wrapping, photos, artwork, and sealing envelopes. They're neat, quick drying, and won't cause paper to wrinkle. The glue dries quickly and washes out for easy cleanup. The odors produced vary greatly from brand to brand. (Fig. 5-7)

Used on paper, glue sticks create a permanent bond. On fabric, they're used only for basting and can be sewn through without leaving any residue on the needle.

Major brands include: Aleene's, Dennison, Dritz, Devcon *Duco,* Elmer's, Kony Bond, LePage's, Loctite

Desk Set, Pritt, 3M Scotch, Ross, and Eberhard Faber *UHU.* You also can find colored glue sticks, such as *UHU Stic Color* or Elmer's *School Glue Stick,* that go on purple or blue (so that you can see where they're positioned) and then dry clear.

Glue stick formulas in general are quite similar. Compare prices for the best buys. You may find that some glue sticks are soft and leave more glue than necessary. Others go on very smoothly, while still others are stiff and require more pressure to apply. One major factor is the age of the glue—as they get old, glue sticks will dry out and eventually become impossible to apply smoothly. Neither chilling nor warming in a microwave will solve that problem.

In general, a glue stick need not be chilled in the refrigerator before use, as some crafters suggest. But storing it in a cool place can slow its tendency to dry out if you live in a warm, dry climate.

CLEAR PAPER GLUES

Most of the clear craft glues on the market are listed on page 19. A few others are marketed specifically for working with paper, but these generally have the same characteristics as those already discussed and are an alternative to white glues (see Chapter 2). Brand names include: Loctite *Desk Set Clear Gel Glue* and Ross *Paper Fix.*

REPOSITIONABLE PAPER GLUES

Most of the repositionable appliqué glues listed on page 38 will also work well on paper. Some, such

PROJECT: Personalized Note Cards

Personal stationery speaks a wordless message on behalf of the sender. Use the finest papers and elegant images to create cards for a fraction of designer prices. (Fig. 5-8)

Fig. 5-8

MATERIALS NEEDED

· Plain cards and envelopes (found in discount office and paper outlets or stationery stores)
· Gift wrap, recycled cards, lightweight print fabric, or other interesting paper with distinct and separate motifs
· Gold or silver paint marking pen
· Rubber cement
· At least 1 yard of coordinating ribbon

HOW TOS

1. Cut out decorative motifs from the paper or fabric—either precisely along the edge of the motif or framed in a square or rectangle of the background. Or, using a stencil, cut out an initial from a lovely textured paper. (Fig. 5-9)

Fig. 5-9

2. Place each motif on a card and lightly trace around it with the marking pen.
3. Apply rubber cement to the back of each motif and to the area inside the traced markings, and let it dry.
4. Carefully position the motifs over the marked areas. Lay a sheet of white paper over each motif and burnish it down as described on page 58. If you've missed a little on your positioning, use the marker to redraw any covered lines.

as *Sticky Stuff,* are permanent on paper. Others, such as *Press n Peel,* can be used as a repositionable. In addition, there are a number of lighterweight glues developed and marketed specifically for repositioning paper. (Fig. 5-10)

In general, the repositionable paper glues are nontoxic, wash off easily, and have sponge applicator tips (many with a chisel-point for better accuracy). Some may stain fabric, so test before using them directly on surfaces other than paper.

Brand names in this category include: Kony Bond *2 in 1 Glue Marker* and *School Glue Marker,*

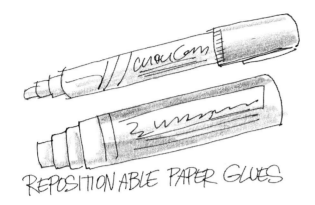

REPOSITIONABLE PAPER GLUES

Fig. 5-10

Dab Stic, Gonicoll *Stick and Peel Glue,* Loctite *Desk Set Ball Point Glue Pen* (with a unique ballpoint pen tip and "pump cap" to prevent clogging), Uchida *Marvy Glue Marker,* and *Zig 2 Way Changin' Glue* (in two sizes).

With any repositionable glue, you can simply apply it and let it dry for a temporary bond. The item you apply it to can be positioned on any surface and peeled off a number of times without leaving any residue. This works well for wall, fabric, and quilting stencils, rubber-stamp masking, stickers, or any project that requires a temporary bond. These glues also work well for small foiling jobs—just rub on the foil after the glue dries, following the instructions on page 50.

"I like to coat the back of a cardboard shape with repositionable glue like Zig 2 Way, then use it for a temporary template for machine quilting. The glue makes it easy to move the template around on the fabric without having to retrace the shape. I never leave the template on overnight, storing it on plastic-coated freezer paper."

Robbie Fanning, *co-author, The Complete Book of Machine Quilting, Second Edition*

For a permanent bond, position the surfaces together while the glue is still wet. Particularly popular with rubber stampers, you can add flocking, embossing powder, confetti, or glitter to any wet glue. The glue also works well for photos, envelopes, labels, gift wrapping, collages, or any other paper project that requires a permanent bond.

Several products on the market, including those from Zig, Kony Bond, and Loctite, make the job of deciding when the glue is dry easy. They go on blue, so you can see the position more accurately, and are completely clear when dry.

SPRAY ADHESIVES

You can attach paper to practically any surface, using a spray adhesive. It's strong, dries clear and quickly, and will not bleed through, wrinkle, or curl paper. Use paint thinner, turpentine, or denatured alcohol for cleanup. (Fig. 5-11)

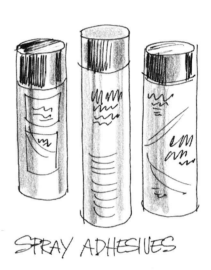

SPRAY ADHESIVES

Fig. 5-11

Use spray adhesive for a wide range of projects, including collage, mosaics, sand and seed painting, découpage, applying glitter or flocking, stenciling, and mounting.

All-Purpose Sprays

When using a spray adhesive, the bond can be either permanent or repositionable, depending on the application method you use. Follow these general guidelines:

▲ Be sure your work area is well ventilated and shake the can well before beginning.

▲ To avoid getting spray on the surfaces around your project, turn a large cardboard box on its side and place the object inside when spraying. (Fig. 5-12)

Crafting with Paper **61**

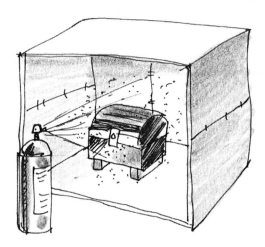

Fig. 5-12

▲ Hold the spray can about 8 to 12 inches from the project surface (following the manufacturer's instructions) and move the can back and forth, slightly lapping each previously sprayed area and applying a uniform coat.

For a repositionable bond, apply a light coat to one surface. After a few seconds for drying, position the sprayed item. Carefully peel it off if you need to reposition it.

For a permanent bond, apply a medium spray coat to both surfaces. Allow the adhesive to dry partially before bonding—the less porous the surface, the dryer the adhesive can be.

When finished spraying, turn the can upside down and press the nozzle until no more adhesive escapes. Doing this will prevent clogging.

You'll find a wide range of all-purpose spray adhesives on the market. Some are made specifically for artists or for home decorating, hobbies, and crafts. Others are for general use. Brand names include: Blair, Bond, Creatively Yours, DAP, Devcon Duco, Duro, Elmer's, Krylon, Scotch, Sprayway, and 3M.

Repositionable Sprays

Although any repositionable spray adhesive can be applied with a temporary bond, several have been formulated as pressure-sensitive sprays. Simply apply a light, even coat, following the manufacturer's instructions and the tips given previously.

Wait a few minutes for the adhesive to set up, then place it on a scrap of fabric to remove any excess residue before positioning it, using firm, even pressure. When stenciling, the paint or glue will stay precisely within the cutout area for a sharp image.

After several repositionings, the adhesive may lose its effectiveness, so repeat the application process. Temporarily store the stencil on waxed paper. For longer storage periods, clean off the adhesive first.

Several brand names of repositionable spray adhesives are: Blair *Stencil-Stik,* Creatively Yours *Stencil Adhesive,* Delta *Stencil Magic,* and Krylon *Easy Tack.*

PROJECT: Super Simple Glittered Ornament

Make beautiful Christmas ornaments as easy as 1-2-3. There's minimal cleanup because the adhesive and glitter go directly into the bulb. (Fig. 5-13)

MATERIALS NEEDED

· A clear glass ornament with a removable top
· Fine glitter
· Spray adhesive
· ¾ yard of narrow ribbon

HOW TOS

1. Remove the ornament top and spray adhesive into the ornament, holding the nozzle right at the opening. Place a paper towel over the opening and shake well.
2. Pour glitter into the opening, cover it again, and shake again. Then wait five minutes and shake out any unattached glitter.
3. Replace the ornament top and tie on a decorative bow.

Fig. 5-13

6

Power Bonding

All of the multipurpose glues and adhesives discussed in Chapter 2 are used for bonding porous and semiporous materials. But bonding nonporous materials (those with a hard, smooth surface), especially to other nonporous materials, is a more difficult job because the adhesive has much less to cling to when the surface is hard and smooth.

In this chapter on power bonding, you'll learn about the multipurpose adhesives developed for strongly bonding nonporous surfaces. The products are practically all solvent-based and are either water-resistant or waterproof, although a few are not recommended for long-term immersion in water. Always read the labels.

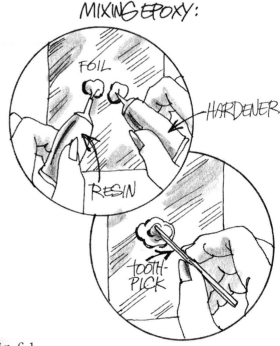

Fig. 6-1

EPOXIES

One of the strongest adhesives available for crafting projects, epoxy comes in two parts (a resin and a hardener) and must be mixed together thoroughly, usually in a 1-to-1 ratio, before it will harden and gain its strong adhesive properties.

Dried epoxy can be painted, sanded, and drilled. Paint pigments can be added during mixing, so you can use it to repair small chips in dinnerware or other ceramics. Most epoxies will withstand heat up to 250 degrees, so they're dishwasher safe.

Use epoxy to glue any hard surface—such as metal, glass, wood, ceramics, porcelain, hard plastics, and fiberglass—to any other hard surface. It's not recommended for polyethylene (clear lightweight plastic), polypropylene, plastic foam, or soft wood (because it will be stronger than the wood, it will have no give and can cause splitting).

Mix epoxy on a disposable surface—a piece of foil works well. Stir and apply it with a toothpick or other disposable applicator. If the surfaces you're joining need to be cleaned, use acetone or an acetone-based nail polish remover. (Fig. 6-1)

Setting times vary from as little as 5 minutes to as much as 30 minutes (read the instructions), and a hair dryer can speed up the process. Use a fast-setting epoxy for smaller jobs and a slower-setting epoxy for large areas or when you need more time for positioning. You must use any mixed epoxy before it sets, so mix only what you'll need. To hold two surfaces in position until the epoxy sets, use cellophane tape.

Wait until the epoxy has fully cured before using the glued project—curing can take 24 hours or longer. Removing epoxy after it has cured is difficult, but solvents are available to do this. One is *Attack*, a methylene chloride from Hughes Associates. Soak the project in it overnight and then clean it with acetone. This is powerful stuff, so use extreme caution!

Some types of packaging make the mixing and application of epoxy easier. For small jobs, you'll find easy-mix pouches that let you squeeze one part into the other (by breaking a thin seal) and mix before opening—no guesswork or mess, and you can apply it right out of the pouch onto the project. Brand names for this type include: Creatively Yours *Quick Set Epoxy Pouches*, Loctite *Poxy Pouches*, and 3M *Metal & Ceramic Adhesive*.

Epoxies are also packaged in side-by-side tubes with a single plunger (to ensure that you use equal amounts) or in separate tubes or squeeze bottles. A few of the many brands on the market include: Devcon *2-Ton Epoxy* and *5 Minute Epoxy*, Duro *Master Mend*, Elmer's *Epoxy* and *Super-Fast Epoxy*, and Ross *Epoxy Adhesive* and *Fast Set Epoxy Adhesive*.

Because of its strength and gap-filling ability, epoxy is often used for assembling jewelry. Metal findings (such as pin and earring backs), gems, buttons, and other nonporous surfaces will all bond securely. Later in this chapter you will find other adhesives used for jewelry making, but none is stronger than epoxy.

You'll need to apply the epoxy to only one of the surfaces. When making jewelry, wait overnight before wearing the piece to be sure that the epoxy has cured completely.

ACRYLIC ADHESIVE

I recently discovered another two-part crafting adhesive that's not an epoxy but is acrylic based. One product marketed by Halcraft USA, *Hal-Tech 2001*, is ideal for jewelry making, especially when layering several pieces in the same project. (Fig. 6-2) Loctite *Depend II* adhesive works in the same way.

To use acrylic adhesive, brush the activator onto both surfaces being joined. Then put the adhesive on only one surface and join the two pieces together. You'll have about 20 seconds for repositioning, and the adhesive will set in 30 seconds. There's hardly any waiting for each part to set before layering another over it, and you don't have to wait overnight before wearing.

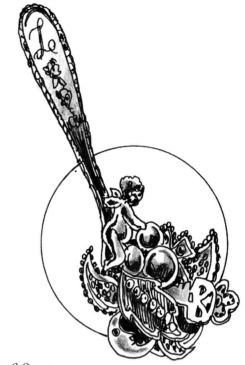

Fig. 6-2

PROJECT: Classic Button Earrings

With very little effort, you can make interesting earrings to match any outfit. Your options are limited only by the pairs of buttons you can collect.

Fig. 6-3

MATERIALS NEEDED

· Two matching buttons—the size, shape, and color you want the earrings to be
· Two clip-on or post earring backs (for larger, heavier buttons, clip-ons usually work best)
· Epoxy
· Wire cutters
· Metalworking file

HOW TOS

1. If the buttons have metal or plastic shanks, clip them off as close to the button as possible using the wire cutters.
2. File away any remaining part of the shanks to make the backs as smooth and even as possible.
3. Mix the epoxy, immediately apply it to the earring backs, and position the backs on the buttons until the epoxy sets. If the earrings are perfectly flat, the backs will remain in place without sliding to one side. If they are not flat, you'll need to prop the earrings so they won't tilt, or tape the backs into position until the epoxy sets.

Acrylic adhesive works on metal, glass, ceramic, marble, crystal, hard wood, stone, and some plastics. Other acrylic adhesives have been available for some time, but only for industrial uses.

SUPER GLUES

Anytime you hear a product referred to as a "super glue," it falls into an adhesive type called cyanoacrylates. These glues are relatively expensive but take only a drop or two to bond strongly. Liquid super glues work best on hard, nonporous surfaces, so I've included them in this chapter. Super glue gels (see Chapter 2) work best for porous and semiporous surfaces. (Fig. 6-4)

One disadvantage of super glue is that you have very little time to position your project surfaces accurately—you need to be right the first time. Super glue is also quite runny and it bonds practically everything it touches (even skin) before you know what's happened. Wear gloves when using it, or be **very** careful. ***Note:*** Look for technical progress in this area. Borden now markets *Krafty Glue LSA* ("less skin adhesion"), which features delayed bonding to skin.

If you do have an accident with super glue, solvents developed especially for super glue are available. Keep one on hand if you'll be doing much super-glue work. Another removal option is acetone or acetone-based nail polish remover, but you'll need a lot of patience—it takes a while. (Fig. 6-5)

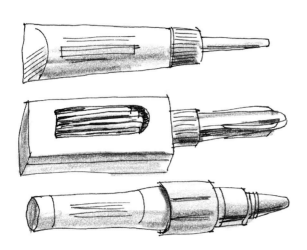

Fig. 6-4

Fig. 6-5

Most often used for model-making, attaching light florals to decorative bases, or other small-scale jobs, super glues are a very strong (but not the most durable) adhesive. You'll also find them handy for quick repair jobs on rigid, nonporous surfaces. The few materials super glue won't bond or work well with are *Teflon,* polyethylene, polypropylene, and foam rubber.

A super glue's outstanding characteristic is its fast bond, which has its advantages and disadvantages. You'll need to wait only a few seconds for super glue to set, so clamping won't be necessary and you can complete your project quickly. The glue will also penetrate and help seal hairline cracks.

Another related product on the market is super-glue activator. This reduces the setting time and can be used in combination with super glue to build up and smooth out an uneven surface before bonding. Read the manufacturer's instructions because they vary from brand to brand.

You can extend the life of super glue by storing it in the refrigerator to keep it from drying out. But when you remove it, let the glue warm up to room temperature before opening the cap.

Recently some companies have developed a "new" type of super glue that will bond to more types of surfaces. It's still a cyanoacrylate, but the new glue has a viscosity somewhere between a liquid and a gel. In fact, there isn't much difference in the content of any of the super glues, but the containers they're packaged in can vary.

All super-glue containers have some type of small tip for pinpoint application, but you can also find push-button applicators and special pens, which will give you better control. In addition to Borden *Krazy Glue,* which is one of the most widely recognized brands, super glues are sold under many brand names, including: Bond, Creatively Yours, DAP, Devcon, Duro, INTAC, Kony Bond, Loctite, Ross, 3M, Scotch, Surehold, and Quicktite.

CLEAR CEMENTS

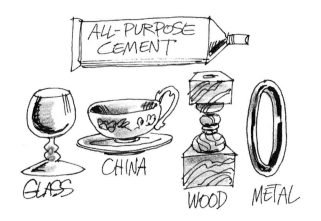

Fig. 6-6

You'll find many clear cements available to help with a variety of crafting projects and general household repairs. One cement that works especially well for jewelry is G-S Hypo-Tube Cement from Germanow-Simon. It's used in the watch repair industry and by optical professionals. Because it's not as fast drying as super glue, you have a little more time for aligning and positioning, and the tiny applicator tip gives you a precise amount for neat, clean work.

"When stringing beads, seal the knots with G-S Hypo-Tube Cement *because it penetrates the thread exceptionally well."*

Yvonne Perez-Collins, *Guide to Beading By Hand and Machine*

All-Purpose Cements

The formulas for many of the all-purpose cements on the market vary from brand to brand, so always read the recommended uses and characteristics of each. A good all-purpose cement can help you complete a variety of crafting and repair projects successfully. (Fig. 6-6)

In general, all-purpose cements are strong, but they are not the strongest adhesive available. They have greater holding power than white glues (see Chapter 2) but less than epoxies or many of the high-tech adhesives. They have a slower setting time and drying time than most other adhesives in this chapter, but they still dry faster than a water-based glue. When dry, all-purpose cements are invisible, water-resistant, and unaffected by weather and temperature changes. Some are also dishwasher safe.

Suggested uses for all-purpose cement vary from brand to brand. Common uses include model-making and work on glass, china, ceramics, porcelain, wood, and metal. Other uses listed on only some brands are for canvas, rubber, gemstones, shells, tile, leather, and most plastics. With few exceptions, all-purpose cement will melt plastic foam.

Most formulas of clear cement can be removed with acetone or acetone-based nail polish remover. Brand names include: Barge *All-Purpose Cement,* Bond *527 Multi Purpose Cement* and *Victory Household Cement,* Creatively Yours *Crafter's Cement,* Devcon *Duco Cement* and *Weldit Cement,* Duro *Household Cement,* Elmer's *Household Cement,* Ross *Household Cement,* 3M *Super Strength Adhesive,* and Scotch *Super Strength Adhesive.*

Contact Cements

Although not often used for crafting, contact cements may be helpful for some special jobs. They're usually used for laminating two larger surfaces together on contact so that clamping isn't necessary, such as bonding other materials to wood. (Fig. 6-7)

To apply a contact cement, spread it on both surfaces and let it dry (or let it become tacky, depending on the brand). When the two surfaces are joined together, they will bond instantly, allowing no time at all for positioning or adjustment. And the bond usually becomes even stronger with age.

Contact cements vary from brand to brand, but they all produce a strong bond. Some are solvent-based, but many are water-based, nontoxic, and easier to clean up—check the label. The water-based cements usually take longer to set.

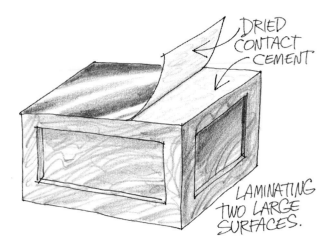

DRIED CONTACT CEMENT

LAMINATING TWO LARGE SURFACES.

Fig. 6-7

Contact cement brands include: Elmer's, Devcon, Duro, Ross, Weldwood, and Wilhold.

HIGH-TECH ADHESIVES

On page 22, I discussed the high-tech adhesives available today for many different crafting jobs. They're strong and fast and work well on all types of surfaces. For example, *Quick Grab* gained a good reputation for working well on dollhouse projects.

High-tech adhesives generally can be used on the same surfaces as an all-purpose cement, and many will melt plastic foam. They're also waterproof and dishwasher safe.

Super-creative author Jackie Dodson tells me that she would be "lost without Goop." *Household Goop* is her favorite, because it is inexpensive and available at her local hardware store. One of her bright ideas is for making button jewelry without cutting off the shank backs of the buttons. She squeezes out a big glob of the glue where she wants it and simply pushes the button shanks into it. The buttons end up at different heights (because of the size of the shanks) and create added interest in a multibutton arrangement. If she wants to reuse one of the buttons, she can simply pry it off.

Jackie also sends another great tip from a recent book:

"One of the pillows we featured uses expensive decorative buttons that we wanted to remove during laundering. So we used Goop to glue the buttons to removable button covers. If we ever want to use the buttons for something else later, we can pop them off the button cover hinges."

Jackie Dodson and Jan Saunders, *Sew and Serge Pillows, Pillows, Pillows*

PROJECT: Distinctive Dominoes Pin

Raid your attic or storage closet for that old dominoes game that nobody plays anymore. You can turn the dominoes into whimsical costume jewelry. (Fig. 6-8)

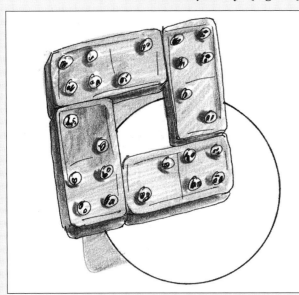

Fig. 6-8

MATERIALS NEEDED

- 4 dominoes—use any combination of number dots you wish
- Small rhinestones or other gemstones—one for each domino dot
- One pin back
- High-tech adhesive

HOW TOS

1. Lay out the dominoes so that the dots form a pleasing design (see the illustration above).
2. Working on a small foil-covered surface, squeeze out a small puddle of the adhesive. Apply the adhesive, using a toothpick, to glue the four adjoining areas. Hold the pieces firmly in place until the adhesive sets (this should take a few minutes or less—read the instructions).
3. Again using the toothpick, apply a small spot of adhesive and one rhinestone to each dot (see Fig. 4-2).
4. After the adhesive sets, turn the pin over and glue the pin back at an angle across one corner.

Rx for Special Jobs

- ▲ Wood Glues
- ▲ Leather Glues
- ▲ Glass Adhesives
- ▲ Plastic Adhesives
- ▲ Foam Glues
- ▲ Reusable Adhesives
- ▲ Miscellaneous Specialty Products

Walk through any builders' supply, hardware, or auto-parts store and you'll spot many more adhesive products than I have space to mention in this book. Some glues are specifically for building and maintaining boats, others are only for automotive work, and still others are for plumbing and household repair. The list is practically endless.

Many of these job-specific glues and adhesives can be helpful only if you're working on an unusual crafting project, but several glues are applicable to general crafting that I haven't already mentioned.

WOOD GLUES

Because this is a book on all types of crafting, it's not possible to discuss all the details of fine carpentry and woodworking—those are specialized, highly skilled pursuits. Instead, I'll simply list the types of products available and recommend crafting applications.

Elsewhere in the book, I've discussed a number of glues that can be used for bonding other materials to wood. In this section, I will discuss only those products recommended specifically for wood-to-wood bonding. Dollhouses, clocks, birdhouses, and boxes are popular woodcrafting projects, as well as picture framing and assembling wood cutouts. (Fig. 7-1)

In general, wood glues are slower drying than craft glues (to allow more time for exact position-

ing), so clamping is usually required. Any woodworking supply store or mail-order catalog can provide you with the specific type of clamps you'll need for your particular job—they vary considerably.

When clamping, try to maintain a firm, even pressure and follow these guidelines:

▲ Avoid gluing green or uncured wood. The joint probably won't hold if the wood isn't dry first.

▲ Position wood blocks or dowels between the clamps and the surface to spread the load on large or difficult areas.

▲ Use a pad or other soft material to protect the surface under the clamps.

▲ Apply pressure gradually, checking the alignment as you go.

▲ Don't use excessive force—that will force glue out of the joint and weaken the bond.

Most wood glues are water-based and nontoxic. Always read the label and follow the manufacturer's instructions because the glues vary greatly. Don't continue work on a project until the glue has completely cured. High temperatures can shorten the curing time, while low temperatures and high humidity will lengthen it.

On any gluing project, you must select the correct adhesive to get a good bond. This is also true with wood glues—they must work under the temperature, moisture, and stress conditions to which they're subjected and for the type of wood you're using. They also must be applied and clamped correctly. In most cases, use a brush for a smooth application and complete coverage. For larger or multiple jobs, you may prefer a special applicator container for rolling, injecting, or brushing the glue. (Fig.7-2)

Perhaps the most important factor in effective wood-glue application is adequate surface preparation. Before gluing, strip or sand the wood to be sure that all paint, wax, or varnish is removed.

Surfaces may need to be smoothed or roughened, depending on the wood and type of product you're using. Oily wood surfaces may need to be wiped down with a volatile solvent such as naphtha or acetone (and then dried) just before applying the glue.

Fig. 7-1

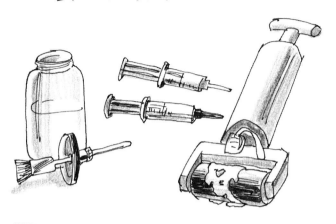

Fig. 7-2

Fig. 7-3

White Glues

The white glues sometimes used in woodcrafting are the same polyvinyl acetate (PVA) glues discussed under white glues and tacky glues, beginning on page 12. These glues should be used only for projects that will undergo little stress during use and when the finished item will not be exposed to water or dampness.

Refer to the specific use recommendations and the brand names listed in Chapter 2. Other white glues marketed specifically for woodworking include: Delta *Woodwiz*, Devcon *Gripwood White Glue*, and Franklin *White Glue*.

When using white glue for woodworking, the setting time will be at least one hour and most brands recommend leaving the project clamped for at least 24 hours, until it completely cures.

Aliphatic Resins (Yellow Wood Glues)

The most common type of woodcrafting glue used today is a white glue mixed with a resin to produce a stronger bond and a faster setting time—30 minutes or less. It works well on both soft and hard woods. (Fig. 7-3)

Like white glues, yellow wood glues are non-toxic and odorless. They clean up with warm water before drying. Never use these glues on outdoor furniture or on anything that will be exposed to moisture. Instead of drying clear, like white glues, yellow glues will probably retain some of their yellow color when dry. Others are specially formulated to dry darker for use with darker woods.

There are several advantages in using a yellow wood glue instead of other wood glues. It has good heat resistance and is not affected by solvents from paint or finishes. When used at cooler temperatures (down to 45°F), the glue will still form a good bond, although the curing time will be longer. Yellow wood glue comes in an easy-to-use liquid form, and, probably most important, it can be sanded after drying.

Some popular brand names include: Aleene's Professional Wood Glue, Elmer's Carpenter's Wood Glue, Franklin International Titebond, Le-Page's Original Strength Wood Glue, Loctite Wood Worx, and Ross Professional Wood Glue.

Garrett Wade, one of the larger woodworking mail-order catalogs, offers several white glue (PVA) and resin emulsions with specific characteristics. Its Yellow Woodworkers Glue is for general-purpose wood work, while 202GF Gap Filling Glue works especially well for furniture making. Slo-Set Glue allows more time for complicated assembly work, and Special Laminating Glue will reduce or eliminate bleed-through on veneers. Today there seems to be a special glue for just about any job.

Waterproof and Moisture-Resistant Wood Glues

Waterproof and moisture-resistant glues have been available for many years (requiring premixing and/or heat application), but now you can use a premixed, liquid glue for many of the same jobs. Elmer's *Weather-Tite Wood Glue*, Franklin International *Titebond II,* 3M Brand *Wood Glue* are three examples.

Use this type of glue for birdhouses, outdoor furniture, and any other small woodcrafting project that will be exposed to the elements. (Fig. 7-4)

Fig. 7-4

A popular recent addition to the woodworking market is *Gorilla Glue,* a strong premixed all-purpose glue that stays flexible and is waterproof. It's stable under a wide range of temperatures, won't react with finishes, and is relatively fast-setting.

For the purist or for large jobs, the old favorites are still good choices:

Epoxy. Industrial-grade epoxies are strong and can be completely waterproof. As with the epoxies discussed beginning on page 66, these adhesives must be mixed just prior to use because they set quickly. Wear gloves and follow the recommendations in Chapter 6. Industrial epoxies are often used for laminating veneer, although special hot-melt sheets are also available for that purpose.

Resorcinol. Long recognized as the waterproof wood glue, resorcinol is another two-part adhesive that requires mixing before use. Its great strength makes it excellent for boats, outdoor furniture, and any other project that will be subjected to moisture. (Fig. 7-5)

Resorcinol also withstands most solvents, acids, molds, and fungi. Apply it at temperatures above 70 degrees, clamp for at least 24 hours, and clean up with warm water **before** it sets. Brands include: DAP *Weldwood,* Wilhold, and U.S. Plywood.

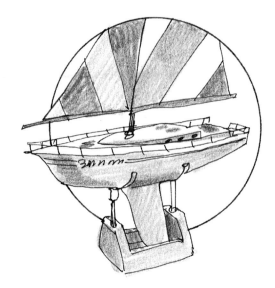

Fig. 7-5

Urea-Formaldehyde. In long-time use for woodworking, this glue has a plastic resin base, is highly water-resistant (although not totally waterproof), and can be sanded when dry. It comes in powdered form and must be mixed with water before use. It's strong but can be brittle, should be applied above 70 degrees, requires clamping at least for 12 hours. This glue won't work well with oily woods. Use caution before curing and with repeated exposure—it's toxic. Brand names include: Weldwood *Plastic Resin Glue* and Wilhold *Plastic Resin Glue.*

Hide and Fish Glues

The world's original glues were probably natural protein emulsions from animal sources. Some are derived from cattle hides and others, such as LePage's *Original Glue* (still available today in premixed form), are from fish by-products. These natural glues are transparent and can be sanded when dry, are not water-resistant, and clean up with warm water. They work well for indoor furniture and other large projects.

Hide glue must cure at least 8 hours at a temperature of 70 degrees or above, so there's a long assembly time. LePage's *Original Glue* must be clamped or weighted for at least 48 hours. Both glues are extremely strong when dry.

Most hide glues must be mixed in flake form with water and then heated and stirred, but now a liquid form is also available. Franklin International *Hide Glue* is the most commonly used liquid hide glue. This glue sets slower than the mixed form and is not quite as strong, but it's much easier to use.

One additional crafting use for hide glue (easier in the liquid form) is for giving a crackling effect to almost any surface. It can be used in the production of carousel horses and other projects that require an aged or antique look. (Fig. 7-6)

Fig. 7-6

To create this effect easily:

1 Apply a sealer and then two or three coats of acrylic (such as *Gesso*) to the surface, allowing each individual coat to dry.

2 Choose two colors of latex paint—one light and one dark. (The paint *must* be water-based.) Apply one coat of paint to your project and let it dry. That color will show through the cracks.

3 Next, apply a generous, even coat of Franklin *Hide Glue*. Be sure to cover the entire surface. Let the glue dry completely—drying may take up to 12 hours.

4 Apply the second color of latex paint without overlapping the strokes, but, again, be sure to cover the entire surface. Work quickly because the crackling process begins immediately.

5 After the paint dries, apply a waterproof sealer.

Another product used to create an aged effect is Bond *Crackle & Age*.

Specialty Wood Glues

Other glues and adhesives that will work with wood are mentioned throughout this book. Notable examples are super glue gel (Chapter 2) and super glue (see Chapter 6). Satellite City *Hot Stuff* is only one example of a cyanoacrylate (super glue) marketed especially for woodworking. In addition to bonding, it can be used on thin woods before carving to prevent breaking. Two other forms, *Special-T Hot Stuff* (medium viscosity) and *Super-T Hot Stuff* (high viscosity), are helpful when you need a glue with more gap-filling properties. (Fig. 7-7)

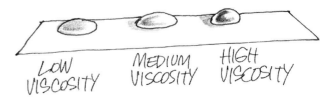

Fig. 7-7

Hot-melt glues (see Chapter 8) also can be used for woodcrafting. They're waterproof, gap-filling, somewhat flexible, and easy to apply, but most glue-stick formulas aren't exceptionally strong. Additional wood glues include:

Casein glues. Another old favorite, casein glues are water-resistant and sandable. They're especially good with oily woods and in cooler (down to 35-degree) work areas. They must be mixed with water before applying. Because they're nontoxic, they're often used for toy construction. Clamping time is only two to three hours.

Specialty cements. All-purpose and contact cements are discussed beginning on page 69 for a variety of surfaces, including wood. Other adhesives in this category were developed for specific wood uses. Bond *Victory Fast Grab* was formulated for bonding wood on dollhouses and miniatures. Testors *Cement for Wood Models* comes in two formulas produced especially for work on balsa wood and other porous materials. The glue must be applied to both surfaces, allowed to dry for some seconds, then pressed together firmly, following the specific label instructions.

A note on model-making with wood: Super glues (page 68), aliphatic resins (page 75), and spray adhesives (page 61) are also commonly used today. Consult your local hobby retailer or a mail-order catalog such as *Tower Hobbies* (see the Resources section at the end of the book).

LEATHER GLUES

Super glue gel and high-tech adhesives (see Chapter 2) and most of the fabric glues discussed in Chapter 3 can be used for gluing leather, suede, and synthetic suede. Always test first. Rubber cement, discussed on page 57, often is used for hemming purposes.

Two other products specifically marketed for use with leather should also be noted: Aleene's *Leather Glue* (**also leather dry-cleanable**) and Tandy *Crafts and Leather Cement*. Both will bond leather to leather as well as leather to other materials, even jewelry findings, but they have different compositions. Read the labels for the exact use instructions.

GLASS ADHESIVES

Many of the glues in Chapters 4 and 6 will work for bonding other materials to glass. Some other glues, however, were developed specifically for glass-to-glass jobs.

Clear Silicone

The major type of adhesive used in crafting to join glass surfaces is clear silicone. It works well because it's actually a very flexible sealant, plus it's waterproof and durable. You can use silicone for sealing and caulking jobs around the house, but it works just as well for crafting. The major disadvantage of silicone is that few materials stick to it after it is dry, including paint, glue, or even more silicone.

The two main crafting uses for silicone are glass construction projects (glass boxes and floral terrariums, for example) and dimensional projects (such as three-dimensional paper work and shadowbox displays). Silicone is usually a good choice for constructing any project that will contain water or be kept outdoors.

In addition to glass, silicone works on wood,

fiberglass, most metals, tile, and painted surfaces. It's thick and viscous and fills in any ridges or gaps, so it can also be used with porous surfaces, such as fabric, silk flowers, shells, and plastic foam.

Kathy Lamancusa, Loctite's well-known designer and consultant, suggests creating papier tole (raised dimensional artwork) using silicone. Simply bond your first print to a surface with spray adhesive. After cutting out individual pieces of a second print, squeeze out a candy-kiss-size glue drop in the center of the area to be dimensionalized. Then place your second cutout print on top of the silicone glue drop. (Fig. 7-8)

Fig. 7-8

Setting varies from 5 to 20 minutes or more, depending on the brand of silicone and the conditions. Drying time is about 24 hours, but complete curing will take longer. Although it can be an eye and skin irritant before drying, silicone is not, technically, toxic. To remove any residue, just trim it away with a razor blade or craft knife.

Silicone and silicone-based brand names include: Creatively Yours *Clear Silicone* (from Loctite), DAP Dow Corning Brand *Silicone,* Devcon *Clear Silicone Rubber,* Duro *Clear Silicone Sealer* and *Stick-With-It,* and Elmer's *Stix-All.*

Note: Another product, Duro *Crystal Clear,* works well for gluing glass, but it is not silicone-based. Instead, it is cured in a vacuum by ultraviolet light.

Stained-Glass Adhesive

Glues are available from large craft stores or mail-order catalogs for other types of glass crafting. Al-

PROJECT: Elegant Book Jacket

Nothing could be easier to make than this attractive paperback book cover. Make several for gifts and make one for yourself as well. (Fig. 7-9)

Fig. 7-9

MATERIALS NEEDED

- One 16″ by 7-½″ rectangle of leather or suede
- Two 10″ by 1″ matching or contrasting leather or suede strips
- One 3″ by 4″ rectangle of contrasting leather or suede (or equivalent scraps)
- Stencils or transfers for the monogram
- Leather glue, such as Aleene's Leather Glue
- Waxed paper

HOW TOS

1. Fold 3″ of both rectangle ends to the underside. (Fig. 7-10)

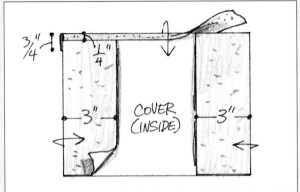

Fig. 7-10

2. Working over a sheet of waxed paper, run a thin line of glue (less than ¼″ wide) between the layered edges at the top and bottom. Weight with heavy books and allow to dry, following the label instructions. (Also put a sheet of waxed paper between the project and the books to protect them from any excess glue.)
3. Brush leather glue evenly over the back of both strips. Wrap one each around the top and bottom edges, leaving ¾″ on the outside of the cover and only ¼″ extending to the back side. Place the cover with the outer side up and weight the glued edges again.
4. Using the stencil or transfer, cut out initials for the monogram. (Just barely cut away any markings as you work.)
5. Apply leather glue to the backs of the letters and arrange them on the center front of the outside cover. Shift your books and waxed paper to weight the monogram also until the glue completely dries.

though authentic stained glass is made by soldering lead between the glass sections, you can now replicate the effect with easy-to-use liquid adhesives.

Real Glass *Liquid Leading* is one such product. It's water-based, nontoxic, and bonds any clean surface, such as glass, plastic, wood, and metal. Because it is black, it looks like leading and is often used for Tiffany-style projects and other stained-glass looks. (Fig. 7-11)

Real Glass *Laminating Glue* is a clear, nontoxic adhesive used for laminating stained-glass pieces onto a clear glass backing. The gaps are then filled in with Real Glass *Grout* to produce a permanent window that looks much like real stained glass.

Crushed-Glass Cement

Other products are specifically marketed to bond crushed glass for inlay work on projects such as

Fig. 7-11

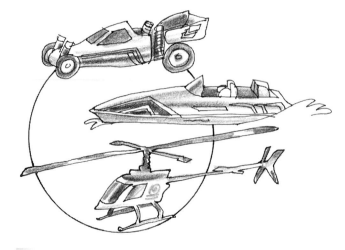

Fig. 7-12

lamp shades, goblets, and jewelry. One example is National Artcraft *Narco Glass Hold,* which is water-soluble and waterproof and dries clear.

PLASTIC ADHESIVES

Many different types of plastics are in use today. Chapter 4 offers suggestions for bonding embellishments to plastic and Chapter 6 gives options for bonding most rigid plastics to each other or to other nonporous surfaces. (Refer to the following section for information on gluing plastic foams.)

When bonding plastic to a dissimilar material, such as wood or metal, where the two will react differently during use, a flexible bond may be required. In that case, silicone (see page 78) is a good option.

For some plastic crafting and for repairs, you may need a strong, gap-filling, water-resistant bond. Also consider using an epoxy for hard plastics (see page 66). One brand of epoxy specifically marketed for plastics is Devcon *Plastic Welder.*

Model-making is one of the major crafting uses for a plastic-to-plastic adhesive. You'll find many model airplanes, cars, trains, and boats that are made of hard polystyrene. (Fig. 7-12)

The main method of joining hard plastic surfaces is called "solvent cementing," in which the solvent softens the surfaces. The surfaces then actually fuse together as the solvent evaporates. Brand names include: Devcon *Duco Plastic & Model Cement,* Elmer's *Model & Hobby Cement,* Ross *Snif Proof Model Cement* (a nontoxic formula), and 3M Brand *Plastic & Model Cement.* (See the suggestions under clear cements, page 69, for cautions and general information.)

A note on model-making using plastics: Epoxies (page 66), super glues (page 68), and several other specialty adhesives are marketed for model-making. For example, the *Top Flight Supreme* line of super glues is one of the best cyanoacrylates available and is sold for this hobby. Consult your local hobby retailer or a mail-order catalog.

FOAM GLUES

Although *Styrofoam* has become practically a generic term, it is a trademark of Dow Chemical Company for its plastic foam product.

The white and tacky glues featured in Chapter 2 work well for bonding to plastic foam. Most spray adhesives will also work with foam for bonding lightweight materials, such as paper, glitter, and other light decorative items. For a faster bond, you can use a low-temperature glue gun (see Chapter 8). (A hot glue gun can also be used, but it will probably melt a small portion of the foam—follow the manufacturer's safety instructions.) (Fig. 7-13)

In addition to the general craft glues used for decorating foam, some are marketed specifically for foam. Bond *Floral Foam Bond* and Modern Miltex *Goo Loo Styrofoam Craft Adhesive* are two that are faster setting than white or tacky glue. Nontoxic Flora Craft *Clear Glue* is another product developed for work with plastic foam.

Because these products vary in formulation, follow the manufacturer's instructions for application and cleanup procedures. To hold embellishments in place while the glue sets, use florists' picks, toothpicks, or short sections of chenille stems.

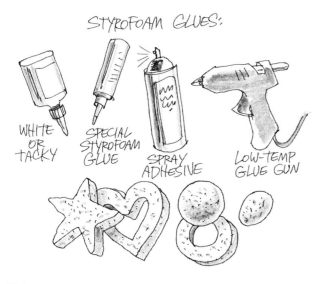

Fig. 7-13

In addition to the glues used for plastic-foam crafting jobs, two aerosols should be mentioned. One is Lomey *FloraLock Stem Adhesive,* which sprays on to hold flowers securely in fresh or dry floral foam, regardless of the temperature or humidity. It sets completely in 45 to 60 minutes to keep the stems from turning and loosening.

Another unique adhesive is Convenience Products *Arrange-It.* This aerosol foam mousse expands to twice its size as it adheres to itself, florals, plastic foam, and other materials. It forms its own feather-weight green foam that in itself is a handy crafting medium. *Arrange-It* dries in about 15 to 30 minutes, giving you enough time to rearrange any decorative materials for the best effect. Use it for swags, cen-

terpieces, wreaths, floral picture and mirror frames, decorative baskets, and many other lightweight crafting projects. (Fig. 7-14)

Also marketed with *Arrange-It* is *Clean-It* (which is necessary for cleaning the valve and craft tip after use) and *Set-It* (a spray adhesive formulated for permanently attaching lightweight decorative items to the foam).

REUSABLE ADHESIVES

Reusable adhesives are plastic, kneadable and adhere to almost any clean, dry surface. And they won't dry out. Use reusable adhesive to hang posters, artwork, and other decorations temporarily and for making the pickup tool described on page 45. These adhesives are also good for securing smaller crafted home decorations from damage by being knocked off a shelf or table accidentally or during an earthquake (something I definitely need in San Francisco). (Fig. 7-15)

Reusable adhesives are safe, nontoxic, and easily removable. Common brand names include:

Fig. 7-14

Fig. 7-15

PROJECT: Victorian Kissing Ball

Combine potpourri, dried and silk flowers, berries, and ribbon to ornament an elaborate pomander. Hang the kissing ball in a doorway or foyer as a fragrant decoration. (Fig. 7-16)

MATERIALS NEEDED

- One rounded plastic foam form
- Dried flowers with small full buds
- Small silk flowers and berries (for accent)
- Dried potpourri
- At least 4 yards of narrow satin ribbon to coordinate (widths can vary)
- One long pin or hat pin
- A plastic foam glue, such as Bond Floral Foam Bond

HOW TOS

1. Turn the form upside down and glue the potpourri in place, beginning at the bottom of the shape and working up to a little past halfway. For easiest application, first coat the foam with glue, then pat on handfuls to cover the surface.
2. After drying, turn the form right side up. Using 1-½ yards of the ribbon, tie a double-loop bow with another loop for the hanger and glue it onto the top of the ball.
3. Glue the dried flowers around the ribbon and cover the top to slightly over the line of the potpourri. Intersperse the silk flowers and berries among the flowers.
4. Tie several bows with 3″ to 4″ tails extending. Run the pin through the center of each bow and then directly into the center bottom of the form, leaving the ribbon tails cascading down from the ball.

Fig. 7-16

Bond *E-Z-Tak*, DAP *Fun-Tak*, Dcor *Prestik*, Devcon *Duco Stik-Tak*, *Elmer's-Tack Adhesive Putty*, Eberhard Faber *UHU HOLDiT*, Loctite *Desk Set Put Ups*, Ross *Tack Tabs*, and Scotch *Heavy Duty Mounting Squares*.

MISCELLANEOUS SPECIALTY PRODUCTS

In this book on crafting with glues and adhesives, some products may not be considered true adhesives but are considered by many crafters to fall into that general area. In addition, a few other job-specific products should be mentioned briefly.

Fabric Stiffeners

On some crafting projects, you'll want to shape or mold material and have it remain set in that position. Fabric stiffeners can be used to do this on doilies, lace, needlework, bows, silk flowers, and other materials. (Fig. 7-17)

Some stiffeners come in a spray form, while

Fig. 7-17

others must be brushed or dipped. Brand names include: Aleene's *Fabric Stiffener and Fabric Draping Liquid,* Beacon *Stiffen Stuff* (pump spray), Bond *Get Set,* Faultless *Instant Microwave Stiffener,* Jurgen *Fabric Drape & Lace Stiffener* (also used for napkin appliqué and papier mâché), Plaid *Fabric Stiffener, Stiffy,* and TAC *Spray Stiff.*

Sheet Adhesives

In Chapter 3, you read about fusible films that need heat for application. These films can be used effectively to bond fabric, but other two-way sticky films are also available for projects in which fusing is not possible.

Crescent *Perfect Mount Film* is repositionable on contact, but it becomes permanent after burnishing. It can be used for mounting photographs, other paper items, and also fabric in arts and crafts projects.

Crescent *Sand Expressions Film* (with a higher tack) is used as a base for sand painting. After affixing the film to a backing, sections can be cut out one at a time so that various colors of sand can be applied without mixing.

If the sand painting is on washable glassware, it can be baked in an oven at 250 degrees Fahrenheit for 30 minutes to make it dishwasher safe. Both Crescent films are acid-free to protect the attached materials.

True Colors International *Stick & Hold* is an all-purpose sheet adhesive that is stretchy and bonds instantly to a wide variety of materials.

Stanislaus Imports markets *Double Stick Adhesive* for clothing embellishment and jewelry making, used with its handmade metallic *Glacé* material. This adhesive is a washable, peel-and-stick product that's also excellent on stretchables such as *Lycra* knits.

Polymer Clay Adhesives

One new product, UHU *Fimo Glue,* was developed for use with polymer clay. Some experts recommend using a high-tech adhesives (see page 22) or one of the powerful glues in Chapter 6.

In a related vein, noted designers of *Friendly Plastic* projects recommend *Crafter's Goop,* also a high-tech adhesive.

Porcelain Setting Agent

With a product called *Petal Porcelain* from Plaid, you can get the look of porcelain in one step. It's used for dipping silk flowers and ribbons to create decorative accessories for wicker, wreaths, and baskets.

Velcro Adhesive

Velcro has many craft applications as an easy-to-use fastener. For some projects, sew-on or iron-on tape works well. For others, you'll find a sticky-back type will give a strong-enough hold.

One other option (when sewing or ironing won't work and a stick-on won't be strong enough) is *Velcro Glue-On Tape* and *Glue-On Adhesive.* It's easy to apply, but you do have to wait 24 hours for it to dry completely before using it.

"Let one hat or basket sport a variety of different looks with changeable ribbon bands whose ends are held together with Velcro Sticky-Back *tape. Cut a length of ribbon to fit snugly around the hat crown or basket rim, adding a 1-½" overlap. Join the ends with the stick-on tape in the overlapped area."*

Ceci Johnson, *Quick & Easy Ways with Ribbon*

Some Like It Hot

- ▲ Glue Guns
- ▲ Glue Pots
- ▲ Microwave Glue

Before glue guns were developed, any crafter who wanted to use hot-melt glues heated small glue "pillows" in an old skillet. We've come a long way since then! Hot glues are used for fast, easy bonding and for an ever-growing number of decorative projects and interesting effects.

Hot glue has traditionally had a resin base with other materials added. Although regular crafting-glue formulas are not particularly strong, the glue is flexible, nontoxic, sets quickly (in about 5 to 15 seconds), and, when kept below its melting temperature, is permanent. Don't use hot glue for outdoor projects, because, although it is not water-soluble, it can be adversely affected by extreme temperature and weather. When the temperature is below freezing or on a hot sunny day (especially if sunlight is magnified through glass), the glue often won't hold.

You can use a hot-melt glue on most surfaces. It fills gaps well and won't seep through most materials. This glue reaches 90 percent of its total bond strength in one minute.

To remove any excess glue after it dries, use a sharp razor blade or craft knife. Hot-melt glue solvents, such as Magic American *Goo Gone,* are widely available to remove dried glue from your glue gun and work area and leave a nonstick coating. (Be sure to read the instructions and precautions before using.) To remove excess glue, place the item in the freezer—the glue will snap off easily when frozen. Of course make sure the item will not be damaged by freezing. On fabric (put a towel underneath), a hot iron will melt most traces of glue if the iron is hotter than the melting point of the glue.

GLUE GUNS

A glue gun can be the quickest and easiest answer on projects where you want to attach embellishments permanently. Glue guns have become not just a convenient method of dispensing glue but also a creative tool.

Originally available only as "hot glue guns," the glue gun of today has benefitted from technological advances and now includes a variety of other models as well. These handy tools melt sticks of specially formulated glue and release a fine glue line when you squeeze the trigger (most glue guns are trigger-fed). (Fig. 8-1)

Today you'll find mini, mid-size (most popular

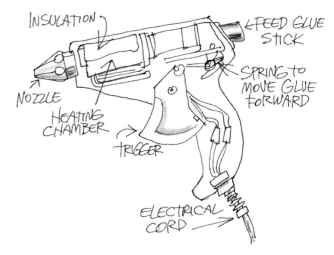

Fig. 8-1

for the average crafter), and full-size glue guns. Some are high (350- to 380-degree Fahrenheit) temperature guns, while others are low-temperature (225 to 240 degrees), and some (called "2-temp" or "dual-temp") can be used for both by changing a heat-setting switch. Some models now are even cordless.

The mini guns are small and often inexpensively priced, sometimes called a "trial size." They're adequate only for small projects and detail work. The main advantage of a mini gun is that the low price allows you to invest in several and use a different color in each gun for multicolor work.

Anyone who does much glue crafting will need a gun that is mid-size or larger. For a few dollars more, you can maintain a continuous flow of hot glue as you complete your project.

Use a hot-temperature glue gun (or setting) when you need more holding power on harder nonporous surfaces, such as metal, hard plastic, glass, and wood. The hot gun is also the best choice when you need a longer working time, when you're working on large areas, and when your finished project will be exposed to heat above 225 degrees Fahrenheit.

A low-temperature glue is good for small, close-in jobs and may be less likely than a larger gun to cause burned fingers. A low-temp gun is a favorite among crafters because it works well for such delicate, heat-sensitive materials as plastic foam, balloons, fabric, paper, dried and silk flowers, ribbon, lace, rhinestones, and beads (thin plastics may be a problem, so test first). It also adheres just as well as a high-temp glue to most materials. Because they use a lower temperature, these guns reach their maximum temperature faster and don't take as long to cool off as high-temps.

The cordless guns, in both low and high temperatures, have a detachable cord or base. They can be used for 5 to 15 minutes, gradually cooling over that time, and then must be reattached and reheated before further use. (Fig. 8-2)

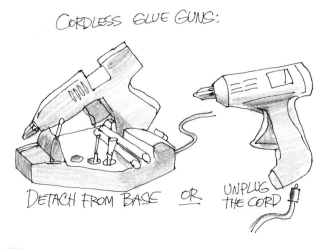

CORDLESS GLUE GUNS:

DETACH FROM BASE OR UNPLUG THE CORD

Fig. 8-2

Respected brand names in glue guns include: Adhesive Technologies, Inc. (including *Crafty Magic Melt*), Arrow Fastener, Black & Decker, FPC Corporation (*Surebonder*), H. B. Fuller, and Stanley. Many generic guns are available, but, remember, you usually get what you pay for. A cheap gun may not heat the glue properly, which can cause excess stringing during application or bond failure later. It can also wear out quickly.

Most of the major glue gun companies have designed different guns for crafters (who are mostly women) than those sold in hardware outlets (mainly purchased by men). The crafting guns often have smaller handles, are easier to squeeze, and sometimes come in "designer" colors.

Some glue guns have been marketed especially for children. They're small, have a lower temperature (under 200 degrees), and include rubber tips and other safety features. The Adhesive Technologies *Stik-A-Roo* line for kids ten and up is one example. But remember, even low-temp glues are hot, so always provide adequate supervision.

A unique new glue gun is the *Floral Pro,* also made by Adhesive Technologies. It is a low-temp gun designed to extrude a much longer line of glue for larger jobs. And instead of a small one- or two-finger trigger, it has a longer, full-finger grip that's more comfortable to use for extended periods.

Another novel product, the *ONEDERGUN,* made by Tecnocraft, features multiple cartridges for quick changes of color or glue type. Both the base holding the cartridges and the gun itself plug in, so the various glues can be kept hot and ready to use at any time during your work. Other features include adapters for smaller-sized glue sticks, specialty applicator tips, and an optional soldering-gun cartridge.

Follow these tips for success in any of your glue gun projects:

▲ Carefully read the instructions for your brand and model before using your glue gun and be sure to follow all the safety tips. Unplug the gun after you're finished and, before putting the gun away, let it cool completely—at least one hour.

▲ Protect your work surface with a specially designed glue mat (see page 89) or disposable paper before beginning the project.

▲ When it's hot but not in use, rest the gun on a stand, glue pad, or heat-resistant plate or glass dish.

▲ If the gun has its own attached wire stand, use the edge of the work surface to press it in or back out as you work. (Fig. 8-3)

▲ Work slowly and methodically to avoid excess drips and glue strings.

▲ You will have *some* wispy glue strings with practically all gun and glue-stick types— sometimes more than others. To minimize the strings, let up on the trigger before pulling the gun away from the project. Then let the

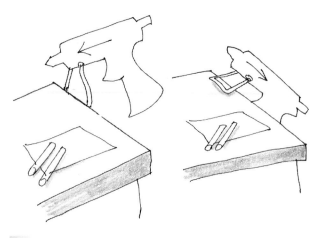

Fig. 8-3

strings that do form dry where they land and just pull them away when dry. Trying to fight them when they're wet is much more difficult.

▲ To keep the gun nozzle clean during your project, have a paper towel handy and wipe off any excess each time before you set the gun down.

▲ To avoid burns, especially with a hot-type gun, apply lotion liberally to your hands before beginning a project. If you will be using one finger to position beads or other small items, wrap it with adhesive tape or a bandage strip first. Another option for applying small items is to use surgical tweezers and keep your fingers out of the glue. *Note:* When bonding hard-to-stick surfaces, lotion may create a problem. If so, wash your hands thoroughly and test again.

▲ When applying a large, heavy item, hold it in place for at least one minute to allow the glue to set. Or wire it on first and reinforce the wiring with glue. Begin with the largest embellishment first and add other, smaller ones after each previous one is in place.

▲ When you are finished using the glue gun, simply pull the plug to turn it off. Don't try to pull the leftover glue stick out—you could damage the gun. When you turn it on again, the gun will remelt and use the glue that is left in it.

▲ When changing to a new color or type of glue (see the next section), just extrude the remainder of the current stick onto a piece of paper or a paper towel. Use a regular clear glue stick (or partial stick) to "purge" the pre-

vious color by trigger feeding it completely through the gun.

Glue Stick Options

Glue sticks used in glue guns vary widely, so always follow the manufacturer's recommendation. Most major companies sell high-quality glue sticks specifically designed for their guns. **Remember, all glue sticks are not created equal.** Be careful when buying sticks in bulk—find out what temperature of gun they were formulated for and test before using them on your project.

Glue sticks vary in size, color, and special formulas. (Fig. 8-4) You'll usually need small (¼"-diameter)

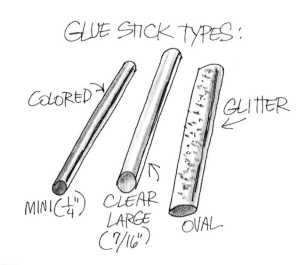

GLUE STICK TYPES:
COLORED
GLITTER
MINI (¼")
CLEAR LARGE (7/16")
OVAL

Fig. 8-4

"mini sticks" for mini guns and regular-size (⁷⁄₁₆"-diameter) sticks for ½" mid-size or full-size guns. Extra-long sticks are also available for larger jobs.

While most glue sticks are round, some low-temp guns require oval sticks (notably the Adhesive Technologies *Crafty Magic Melt* models) as a safety feature and to ensure quality. Low-temp formulas used in a high-temp gun can become too thin or will go through the gun too rapidly and may harm the gun or cause burns.

When glue sticks are purchased in bulk or become separated from their packaging, it may be difficult to tell whether a round glue stick is a low-temp or high-temp formula. To solve the problem, dual-temp sticks (such as Tecnocraft *Hi&Lo* Glue Sticks) have been introduced. They can be used in either type of gun and also are especially handy for dual-temperature models.

Today you can find a wide range of colored glue sticks for decorative work. Primary, pastel, neon, glow-in-the-dark, glitter, and metallic sticks are all available. They come in low-temp, high-temp, and dual-temp formulas and in mini and regular sizes.

In addition to temperature and color, many specialized formulas are also available. In the high-temperature category, you'll find regular and premium all-purpose, crystal-clear, and no-strings varieties. Some types are marketed specifically for wood or plastic. Others have been developed for extra strength, fast set, or slow set. One interesting formula is *Pow'r Stix* from H. B. Fuller. This glue stick allows a longer positioning time (up to five minutes or more), can be used from 380 degrees Fahrenheit to nearly room temperature depending on the application, and can be used on a wide variety of materials, even delicate ones. Always test with your materials before beginning a project.

Special low-temp formulas are also available. Look for a crystal-clear type and for one from Adhesive Technologies developed for use with fabric. It's stronger than other low-temps and can be ironed, machine washed, and machine dried (but not dry-cleaned).

Glue Gun Accessories

Several accessories are available to make crafting with glue guns easier:

Nonstick glue mats. Several companies offer special mats for use with glue guns, including Adhesive Technologies, FPC, and Tecnocraft. Any drips formed while resting your gun on them will peel off when dry. The see-through mats are your best bet because you can also put a pattern under them and trace or fill in the design on top (see the 3-D Sculpting section and Fast Foil Jewelry project following). Rather than a mat, H. B. Fuller offers a transparent silicone release paper in 11″ by 14″ reusable sheets, which provide a large work area.

Special holders. Practically all glue guns come with a built-in wire stand, but many guns are still quite unstable. You can support the handle in a small tray or dish to give it more stability. If you are doing a lot of glue gun work, you might want to use a specially built stand. Stands come in both wood and plastic and usually have a built-in nonstick mat. Some are part of a caddy that can hold glue sticks and other small crafting items right at your fingertips. (Fig. 8-5)

Fig. 8-5

Glue gun nozzles and attachments. Several nozzles are available from some of the major glue gun companies to help you with special jobs. They're generally either long, narrow, or wide to allow an accurate positioning of the glue. FPC also has a cutter that attaches to the tip and cuts off the glue flow quickly after it's dispensed.

Nonstick coating. When you want to keep hot-melt glue from sticking to your hands, molds, or tools, you can use a product such as *Unstick*, which is nongreasy and nontoxic. Another option, suggested by expert Sharion Cox, is to use a mixture of three parts baby oil to one part denatured alcohol in a spray bottle.

Glue Stitching

In 1993, gifted craft designer Maria Filosa developed a unique new technique called "glue stitching." She uses 7-count plastic canvas (in any color), a mini low-temp glue gun, and colored or glitter glue sticks.

Follow these basic guidelines for glue stitching:

1. Using sharp scissors, cut a piece of premium-quality plastic canvas to the exact size you'll need for your project. Use a sharp craft knife to clean any rough edges or slight film from the plastic canvas holes.

2. Working from the underside of the plastic canvas, place the tip of the glue gun directly under a single hole. Squeeze the trigger very gently until glue oozes up through the hole.

3 Extrude enough to create a small "bead" that will rise just slightly above the top surface of the canvas. To finish the bead, release the trigger and scrape the tip of the gun against the canvas underside. (Fig. 8-6)

Fig. 8-6

If possible, use a separate glue gun for each color in your pattern or purge the gun between colors. Begin with the lightest color and work up to the darkest. Half of a clear glue stick will usually clean out the gun when you change to another light or slightly darker medium color. You may need to use one or two whole sticks to purge a darker color.

4 When using only a small amount of color in your project, you may cut the glue stick with scissors and use only a section. But be sure to use at least 1".

Maria offers these trouble-shooting tips for the most successful glue stitching:

▲ Before starting a project, practice making uniform beads on scraps of plastic canvas. Don't be discouraged if your first beads are not the same size or shape. As you continue to practice, the beads will become uniform.

▲ The holes in plastic canvas may vary in size. If you have trouble centering your bead, run the tip of your glue gun around the edge of the hole just before making a bead, slightly coating the edge of the hole with glue. Then center the tip of the glue gun and squeeze out the bead.

▲ You can easily fix a mistake, such as using the wrong color glue or making a bead too small or too large. Wait until the bead has completely cooled. Use a pair of needle-nose

pliers or craft tweezers to grasp the bead at its base and pull the bead off from the front side of your work. Push the pliers through the hole from the front side to loosen any remaining glue, then pull off the residual glue from the underside of your work.

▲ If your glue beads sink back down into the holes, work a few beads at a time and then turn the plastic canvas over so that the beads are on the bottom (gravity will work for you in this case). Wait a few seconds for the glue to set and then continue your work. Sinking usually occurs when the plastic canvas holes are larger than usual or if you are working in a warm area. If your work area is warm, work next to a small fan, which will help cool and set the beads faster.

3-D Sculpting

An entire new crafting medium opened up with the advent of glue sculpting. Glue gun experts Kimberly Nail and Sharion Cox were both instrumental in developing important techniques. Others have continued to expand the possibilities.

One three-dimensional option, pioneered by Kimberly Nail, is glue film:

1 Form a sheet of film by extruding a puddle of colored or glitter glue about 2" in diameter onto a nonstick glue pad.

2 Quickly place a second glue pad over the puddle and use a rolling pin to flatten the glue between the two pads.

3 When cool, peel the pads apart and you have a flexible film that can be cut and shaped. (Fig. 8-9)

Make thicker glue-film layers by pressing with a block or book instead of using the rolling pin. Apply enough even pressure to flatten the glue to about the thickness of cardboard. Cut the thicker film into any shape for many interesting decorative uses.

Sharion Cox has explored and promoted the art of sculpting with a glue gun. The basic glue-sculpting procedure is to extrude glue over a pattern placed under a nonstick pad or surface. (Fig. 8-10) The sculpture can be one- or two-sided and have single or multiple layers. You can contour the sculpture by drizzling on additional thin or thick lines or by layering extra glue in sections to build up texture.

PROJECT: Plastic Canvas Windsock

Try out your new glue-stitching skills on this charming diamond-patterned windsock designed by Maria Filosa. (Fig. 8-7)

MATERIALS NEEDED

- *Uni-Stik* (from Tecnocraft) colored glue sticks—6 neon green, 3 neon red, 2 neon yellow, 4 neon pink, 5 white, and 8 clear
- 7-count black *Plastic Canvas* by Darice: 73 x 25 holes
- ⅝"-wide grosgrain ribbon—2 yards each of hot pink, white, black, and green
- ½ yard of clear thick *Noodles* plastic lacing (from Toner Plastics)

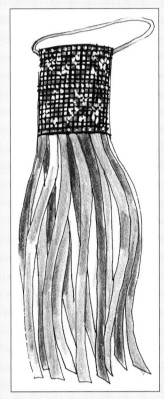

Fig. 8-7

MATERIALS NEEDED *(Continued)*

- Scissors
- A low-temp glue gun
- Needle-nose pliers
- A nonstick glue pad

HOW TOS

1. Following the previous glue-stitching instructions and the pattern illustration (Fig. 8-8), complete the windsock rectangle. There will be three rows of blank spaces along one short end of the plastic canvas.
2. Cut the ribbons into 18" lengths and glue them along one long edge. They will hang side by side from the bottom edge of the windsock.
3. Bring the short canvas ends together, overlapping the three blank rows, and glue to form a cylinder.
4. On the top edge of the windsock, pop out two glue beads directly across from each other, using the needle-nose pliers.
5. Cut the plastic lacing ends at an angle. Place each end in a hole created in step 4, threading from the outside, and knot the ends on the inside of the windsock.

Fig. 8-8

WINDSOCK PATTERN

⊟ WHITE Ⓨ YELLOW
⊠ NEON RED ℙ NEON PINK
Ⓝ NEON GREEN ☐ BLANK

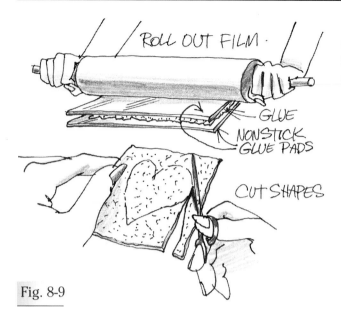

Fig. 8-9

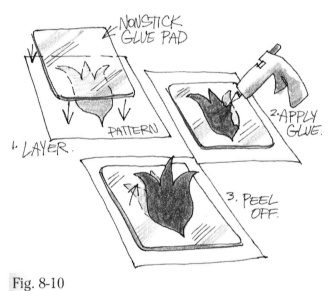

Fig. 8-10

Glue sculpting is now also used to make interesting floral displays, cute kids' ornaments (look for the *Gloobies* pattern booklets by U.S. Star), and other decorative items. (Fig. 8-11)

Fig. 8-11

Hot Glue Jewelry

There are as many great ideas for using hot-melt glue in jewelry making as there are creative designers and crafters out there who enjoy working with it. Pins, earrings, necklaces, and barrettes are all possibilities. (Fig. 8-12)

As a backing for any of your jewelry pieces, use a hardened puddle of extruded glue, a covered mat board or cardboard, buttons, *Friendly Plastic*, or glue film as described above.

Form a fan- or star-shaped glue film using the pattern cut into the bottom of an imitation cut-crystal drinking glass. Moisten the bottom of the glass and squeeze a small puddle of glue in the center. Immediately cover the glue with a glue pad and press with your thumb until it cools. Several of these shapes can be made separately and glued together for an interesting effect. (Fig. 8-13)

To decorate your jewelry, consider using thread, ribbon (metallic or cellophane types are interesting), gems, jewels, sequins, foil, and foldable crafting aluminum. And of course you can use more colored or glitter glue for squiggles, puddles, and outlines.

If the pattern calls for a base, you can make it by placing the dried sculpture vertically into an oval of warm glue until the base hardens. Trim any excess glue with a craft knife and round out the edges with the hot tip of your glue gun. After the project dries, you can remove any excess glue strings with cellophane tape.

Although colored and glitter glue sticks can be used for any glue-sculpting project, you can also use regular sticks and paint the project with any type of water-based acrylic paint after it dries. Use a clear sealer over the paint.

Sharion Cox has written twenty-two Cox Co. pattern booklets featuring glue sculpting (see Resources).

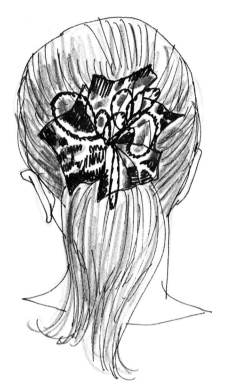

Fig. 8-12

Fig. 8-13

Decorative Effects

Now it's easier than ever to be your own designer. Using a glue gun, you can copy beautiful decorative embellishments on clothing, accessories, and home decor items to express your personal style.

Not only can you use the wide range of colored and glitter glue sticks available, you can also work with uncolored hot-melt glue and decorate it with acrylic paint, sequins, gemstones, and foil.

One of the favorite decorative uses for hot-melt

glues is on garments. Be sure to prewash the garment before applying the glue. Use a transfer for accuracy or try one of the other embellishment ideas featured in Chapter 4. Garments decorated with colored or glitter glue should be hand washed or machine washed on the delicate cycle in warm water. Dry flat—do not use the dryer and do not dry clean. If you're in doubt about the quality of your glue, test first.

Use glue film (see page 90) as an appliqué, cut it into any shape, place it on the fabric, and fuse it to the fabric using a nonstick glue mat as a press cloth. Heat the iron to a rayon or synthetic setting first and press for 10 to 15 seconds. If the design doesn't bond, repeat the process until it does. Overlay different colors of film for added interest. This process also works well for shapes traced over a glue mat and pattern. **Note:** If your fabric is lightweight, such as a T-shirt, use a second nonstick mat under the fabric when pressing to prevent any mishaps if the glue melts through. (Fig. 8-14)

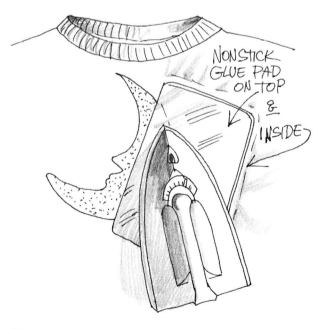

Fig. 8-14

PROJECT: Fantastic Foil Jewelry

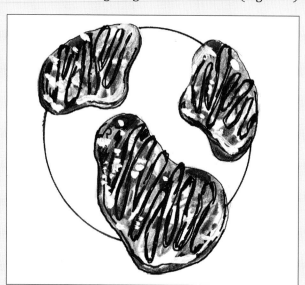

Ceramic and craft designer Kathy Nowicki of South Bend, Indiana, has developed an attractive technique for "solarized" jewelry. Make her simple pin and earrings using a hot-temp glue gun and foil film. (Fig. 8-15)

MATERIALS NEEDED

- A hot glue gun
- Black glue sticks to fit the gun
- Foil film in gold and/or bronze
- Jewelry findings for one pin and two earrings
- A transparent nonstick glue pad

HOW TOS

1. Lay the glue pad over the pattern (Fig. 8-16) and extrude hot glue to fill the shapes. Let them cool and harden completely (at least 30 minutes).
2. Hold the glue gun above each piece and drizzle glue on top of the shapes in a zigzag pattern.
3. Immediately lay foil, shiny side up, over the hot glue. Use your fingertips to pat the foil all around the glue. After the glue cools, pull the foil off to reveal the lustrous design.
4. Drizzle more black glue on top, making more delicate squiggles.
5. After the glue dries completely, apply the pin and earring findings to the backs using more hot glue.

Fig. 8-15

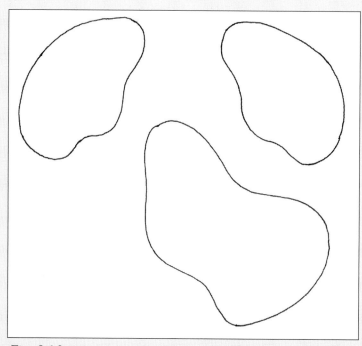

Fig. 8-16

Holiday ornaments are another fun and easy decorative option using a low-temp gun. Turn plastic foam balls into glittery favorites—and they won't break. Revive old Christmas bulbs with swirls or designs of glue.

Artistic author Yvonne Perez-Collins creates center stamen for her beautiful fabric flowers by gluing beads together onto the end of a covered wire, using a low-temp glue gun.

"Using this method, I made scented stamen by gluing together scented beads. I also make stamen with a drop of glitter glue, dipped in water immediately to keep a smooth round shape."

Yvonne Perez-Collins, *Soft Gardens*

GLUE POTS

A hot-melt alternative to glue guns, glue pots free both hands for dipping and placement of your project components. They're great for neatly applying glue to flower stems, pine cones, jewelry findings, and small decorations. And more than one person can use them at the same time. (Fig. 8-17)

Fig. 8-17

One main advantage of a glue pot is that there is little waste. Unused glue can remain in the pot and be reheated later. Use tweezers to dip and position smaller items.

Two models, *Craftmate* (the original) and FPC *Surebonder Glue Skillet,* use high-temp glue pillows. The *Surebonder* can also use high-temp pellets, chips, or sticks. Another model, Adhesive Technologies *Crafty Magic Melt Little Dipper,* uses its oval *Crafty Magic Melt* low-temp sticks. To refill, just push a glue stick into the refill tower.

MICROWAVE GLUE

The world's first microwave-energized glue, *House Works,* was recently introduced by Loctite Corporation. The glue is activated by placing it in a microwave oven on high temperature for four minutes.

The glue becomes hot and easy to use, but the soft foam container and rubber tip stay cool to the touch. Even without being sealed, *House Works* won't dry out between uses. Just heat it in the microwave once more for four minutes and it can be reused again and again.

When you squeeze the container, the glue comes out of the nozzle neatly for precise application and does not form any glue-gun-type strings. It has a strength similar to a dual-temp glue stick and works on many surfaces. Use it for wood, leather, plastic, paper, glass, china, and ceramics, but always test first with your specific project materials to be sure the bond is adequate for the job.

PROJECT: Charming Cup 'o' Flowers

Do you no longer use a favorite mug or cup because it has a big chip in it? Don't throw it away! Turn it into a base for a delightful dried floral arrangement. (Fig. 8-18)

Fig. 8-18

MATERIALS NEEDED

· One cup or mug
· An assortment of dried flowers, some with small full buds
· Long-handled tweezers
· ¾ yard of 1-½"-wide wire-edged ribbon
· Glue pot and glue

HOW TOS

1. To cover the chip or chips and create an interesting arrangement on the side of the cup, clip the stems off several full buds.
2. Using the tweezers, dip the buds in the glue and position them on the cup.
3. Dip the ends of the remaining flowers into the glue and arrange them in the cup.
4. Tie the ribbon in a bow on the handle, trimming the ends diagonally.

Resources

Local craft and hobby stores are usually an excellent source for supplies, information, advice, and inspiration. Mail-order sources are another good source, especially important for those who don't have a craft or hobby retailer nearby.

Key to Abbreviations and Symbols

* = refundable with order
\# = for information, brochure, or catalog
SASE = Self-addressed, stamped (first-class) envelope

MAIL ORDER

Aleene's Creative Living with Crafts, P.O. Box 9500, Buellton, CA 93427, 800/825-3363. Aleene's nontoxic glues, crafting supplies, and how-to books. $2#.

Bolek's Craft Supplies, 330 N. Tuscarawas Ave., Dover, OH 44622, 800/743-2723. A variety of heavily discounted craft supplies and glues. $1.50#. Free freight on orders over $40 to 48 states.

Clotilde, Inc., 2 Sew Smart Way, Stevens Point, WI 54481, 800/772-2891. Sewing notions, related glues (including *Sticky Stuff*), *Magic Glue Wand*, and *Goo Gone.* Free#.

Cox Co., Inc., P.O. Box 1350, Olathe, KS 66061. Booklets with information and patterns on glue sculpting. Free# with SASE.

Craft Catalog, 6095 McNaughten Centre, Columbus, OH 43232, 800/777-1442. A variety of discounted craft supplies and glues. $2#.

Craft King, P.O. Box 90637, Lakeland, FL 33804, 800/769-9494. A variety of discounted craft supplies and a few glues. $2#.

Craftsman Wood Service Co., 1735 W. Cortland Ct., Addison, IL 60101-4280, 800/543-9367. Woodworking supplies, kits, glues, and discounted books. Free#.

Dick Blick, P.O. Box 1267, Galesburg, IL 61401, 800/447-8192. Extensive line of art supplies and glues. Free#.

Enterprise Art, 12333 Enterprise Rd., P.O. Box 2918, Largo, FL 34649, 800/366-2218. Wide range of craft supplies and glues. Free#.

Garrett Wade Company, 161 Avenue of the Americas, New York, NY 10013, 800/221-2942. Woodworking supplies and a variety of glues. Free#.

Jehlor Fantasy Fabrics, 730 Andover Park West, Seattle, WA 98188, 206/575-8250. Specialty fabrics and trims, hot knife and adaptor tips (see page 46), fabric and embellishment glues. $5#.

Leichtung Workshops, 4944 Commerce Parkway, Cleveland, OH 44128, 800/321-6840. Woodworking supplies and glues. Free#.

Nancy's Notions, Ltd., P.O. Box 683, Beaver Dam, WI 53916, 800/833-0690. Sewing and quilting supplies, related fabric and embellishment glues. Free#.

Nasco Arts & Crafts, 901 Janesville Ave., P.O. Box 901, Fort Atkinson, WI 53538-0901, 800/558-9595. Extensive art and craft supply selection, including most types of glue. Free#.

Repcon International Inc., P.O. Box 1345, Nixa, MO 65714. Corrugated cardboard forms for home decorating projects (see page 38). $1# with SASE.

Sax Arts & Crafts, 2405 South Calhoun Rd., P.O. Box 51710, New Berlin, WI 53151, 800/558-6696. Extensive art and craft supply selection, including most types of glue. $5*#.

Sew-Art International, P.O. Box 2725, Spring Valley, CA 91979, 800/231-2787. Sewing notions and accessories, including fusible web. $2*#.

Suncoast Discount Arts & Crafts, 9015 U.S. 19 N., Pinellas Park, FL 34666, 813/572-1600. Wide range of discounted and closeout craft supplies. $2#.

Sunshine Discount Crafts, 12335 62nd St. N., Largo, FL 34643, 813/581-1153. Lower-priced craft supplies, some glues in 200-page catalog. $3# or $5 for priority delivery.

Tandy Leather Co., P.O. Box 791, Ft. Worth, TX 76101, 800/433-5546. Leathers, suedes, related notions, and glues. $3#.

Tower Hobbies, P.O. Box 9078, Champaign, IL 61826-9078, 800/637-4989. Large line of radio-controlled models and accessories. "Tower Talk" (70-page listing of most popular items) Free. Full 300-page catalog $3, or $5 for priority delivery.

Viking Woodcrafts, 1317 8th Street S.E., Waseca, MN 56093, 507/835-8043. Decorative painting and fine art supplies. $10# (*with $100 order).

Woodcraft, 210 Wood County Industrial Park, P.O. Box 1686, Parkersburg, WV 26102, 800/225-1153. Woodworking supplies and glues. Free#.

HELP LINES

3M, 800/3M HELPS.
Ad-Tech, 800/Glue Gun.
Bond, Richard the Chemist, 800/879-0527.
Borden Consumer Response (Elmer's), 614/225-4511 (collect).

Index